IntechOpen

Supporting open minds since 2005

Data Mining - Methods, Applications and Systems
http://dx.doi.org/10.5772/intechopen.87784
Edited by Derya Birant

Contributors
Esma Ergüner Özkoç, Pramod Kumar, Sameer Ambekar, Subarna Roy, Manish Kumar, Makhamisa Senekane, Sana Dardouri, Jalila Sghaier, Selma Tekir, Yalın Baştanlar, Ferda Balcı Ünal, Derya Birant, Elife Ozturk Kiyak, Mehmet Aktas, Oguz Celik, Oya Kalipsiz, Muruvvet Hasanbasoglu, Venkata Subba Reddy Poli, Bouchra Lamrini, Mhlambululi Mafu, Molibeli Benedict Taele

First published in London, United Kingdom, 2021 by IntechOpen
IntechOpen is the global imprint of INTECHOPEN LIMITED, registered in England and Wales, registration number: 11086078, 5 Princes Gate Court, London, SW7 2QJ, United Kingdom
Printed in Croatia

British Library Cataloguing-in-Publication Data
A catalogue record for this book is available from the British Library

Additional hard and PDF copies can be obtained from orders@intechopen.com

Data Mining - Methods, Applications and Systems
Edited by Derya Birant
p. cm.
Print ISBN 978-1-83968-318-3
Online ISBN 978-1-83968-319-0
eBook (PDF) ISBN 978-1-83968-320-6

For EU product safety concerns:
IN TECH d.o.o., Prolaz Marije Krucifikse Kozulić 3, 51000 Rijeka, Croatia, info@intechopen.com or visit our website at intechopen.com.

Meet the editor

Dr. Derya Birant received a BS, MS, and Ph.D. in Computer Engineering from Dokuz Eylul University, Turkey in 2000, 2002, and 2006, respectively. Since 2017, she has been an Associate Professor at the Computer Engineering Department of Dokuz Eylul University. She is also the vice-chair of the department. She was a Visiting Lecturer at South East European University in 2006 and Ege University between 2010 and 2012. Her research interests include data mining and machine learning. She is the author of six book chapters and more than seventy publications. Dr. Birant has supervised more than thirty Ph.D. and MSc students. She has been involved in more than twenty long-term interdisciplinary R&D projects on data mining.

Contents

Preface

Data mining is a branch of computer science that is used to automatically extract meaningful, useful knowledge and previously unknown, hidden, interesting patterns from a large amount of data to support the decision-making process.

This book presents recent theoretical and practical advances in the field of data mining. It reports on a number of data mining methods, including *classification*, *clustering*, and *association rule mining*.

This book brings together many different successful data mining studies in various areas such as health, banking, education, software engineering, animal science, and the environment. The goal is to help data miners, researchers, academics, and scientists who wish to apply data mining techniques in their studies. The references collected in this book may be used as further reading lists.

The main strength of the book is the wealth of the case studies contained within. Chapters cover a number of innovative and recently developed data mining applications. Another important feature of the book is the clear introduction and background information provided at the beginning of each chapter.

The authors of this book have been actively working in the data mining field for years and thus have a lot of experience. They have the skills, knowledge, and expertise needed to share with us about real-world data mining applications. They have aimed at providing readers with a comprehensive understanding of data mining methods and thus present research results in various domains from different points of view. They explain the fundamental data mining techniques for extracting information from a large dataset.

It was not possible for me to accomplish this book without the outstanding contributions of many people. I would like to thank the contributing authors for their excellent works. Much appreciation goes to them for the time and effort they put in. I would also like to thank my husband for his love and support during the editing of this book. I also extend many thanks to Lada Bozic and Marijana Francetic for facilitating administrative matters. Finally, I express my gratitude to the publisher, IntechOpen, for giving me the opportunity to complete this book.

I hope you enjoy reading this book as much as I enjoyed editing it.

Dr. Derya Birant
Associate Professor,
Dokuz Eylül University,
Department of Computer Engineering,
Turkey

Chapter 1

Deep Learning: Exemplar Studies in Natural Language Processing and Computer Vision

Selma Tekir and Yalin Bastanlar

Abstract

Deep learning has become the most popular approach in machine learning in recent years. The reason lies in considerably high accuracies obtained by deep learning methods in many tasks especially with textual and visual data. In fact, natural language processing (NLP) and computer vision are the two research areas that deep learning has demonstrated its impact at utmost level. This chapter will firstly summarize the historical evolution of deep neural networks and their fundamental working principles. After briefly introducing the natural language processing and computer vision research areas, it will explain how exactly deep learning is used to solve the problems in these two areas. Several examples regarding the common tasks of these research areas and some discussion are also provided.

Keywords: deep learning, machine learning, natural language processing, computer vision, transfer learning

1. Introduction

Early approaches of artificial intelligence (AI) have sought solutions through formal representation of knowledge and applying logical inference rules. Later on, with having more data available, machine learning approaches prevailed which have the capability of learning from data. Many successful examples today, such as language translation, are results of this data-driven approach. When compared to other machine learning approaches, deep learning (deep artificial neural networks) has two advantages. It benefits well from vast amount of data—more and more of what we do is recorded every day, and it does not require defining the features to be learned beforehand. As a consequence, in the last decade, we have seen numerous success stories achieved with deep learning approaches especially with textual and visual data.

In this chapter, first a relatively short history of neural networks will be provided, and their main principles will be explained. Then, the chapter will proceed to two parallel paths. The first path treats text data and explains the use of deep learning in the area of natural language processing (NLP). Neural network methods first transformed the core task of language modeling. Neural language models have been introduced, and they superseded n-gram language models. Thus, initially the task of language modeling will be covered. The primary focus of this part will be representation learning, where the main impact of deep learning approaches has been observed. Good dense representations are learned for words, senses, sentences, paragraphs, and documents. These embeddings are proved useful in capturing both syntactic and semantic features. Recent works are able to compute contextual embeddings, which can provide different representations for the same word in different contextual units. Consequently, state-of-the-art embedding methods along with their applications in different NLP tasks will be stated as the use of these pre-trained embeddings in various downstream NLP tasks introduced a substantial performance improvement.

The second path concentrates on visual data. It will introduce the use of deep learning for computer vision research area. In this aim, it will first cover the principles of convolutional neural networks (CNNs)—the fundamental structure while working on images and videos. On a typical CNN architecture, it will explain the main components such as convolutional, pooling, and classification layers. Then, it will go over one of the main tasks of computer vision, namely, image classification. Using several examples of image classification, it will explain several concepts related to training CNNs (regularization, dropout and data augmentation). Lastly, it will provide a discussion on visualizing and understanding the features learned by a CNN. Based on this discussion, it will go through the principles of how and when transfer learning should be applied with a concrete example of real-world four-class classification problem.

2. Historical evolution of neural networks and their fundamental working principles

2.1 Historical evolution of neural networks

Deep neural networks currently provide the best solutions to many problems in computer vision and natural language processing. Although we have been hearing the success news in recent years, artificial neural networks are not a new research area. In 1943, McCulloch and Pitts [1] built a neuron model that sums binary inputs, and outputs 1 if the sum exceeds a certain threshold value, and otherwise outputs 0. They demonstrated that such a neuron can model the basic OR/AND/NOT

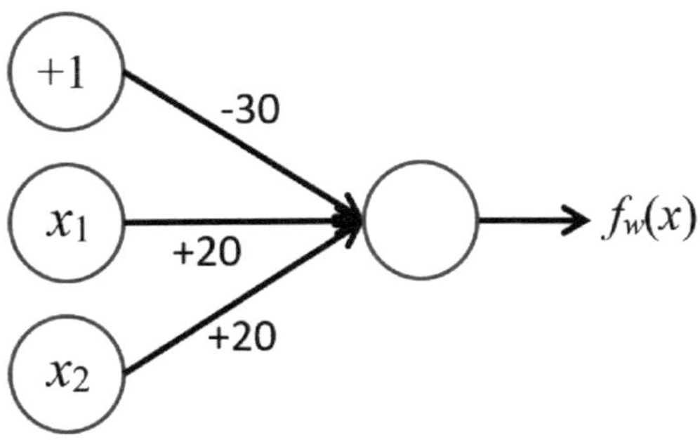

$$f_w(x) = w_0 + w_1x_1 + w_2x_2$$

$$f_w(x) = -30+20x_1+20x_2$$

Figure 1.
A neuron that mimics the behavior of logical AND operator. It multiplies each input (x_1 and x_2) and the bias unit (+1) with a weight and thresholds the sum of these to output 1 if the sum is big enough (similar to our neurons that either fire or not).

functions (**Figure 1**). Such structures are called neurons due to the biological inspiration: inputs (x_i) represent activations from nearby neurons, weights (w_i) represent the synapse strength to nearby neurons, and activation function (f_w) is the cell body, and if the function output is strong enough, it will be sensed by the synapses of nearby neurons.

In 1957, Rosenblatt introduced perceptrons [2]. The idea was not different from the neuron of McCulloch and Pitts, but Rosenblatt came up with a way to make such artificial neurons learn. Given a training set of input-output pairs, weights are increased/decreased depending on the comparison between the perceptron's output and the correct output. Rosenblatt also implemented the idea of the perceptron in custom hardware and showed it could learn to classify simple shapes correctly with 20×20 pixel-like inputs (**Figure 2**).

Marvin Minsky who was the founder of MIT AI Lab and Seymour Papert together wrote a book related to the analysis on the limitations of perceptrons [4]. In this book, as an approach of AI, perceptrons were thought to have a dead end. A single layer of neurons was not enough to solve complicated problems, and Rosenblatt's learning algorithm did not work for multiple layers. This conclusion caused a declining period for the funding and publications on AI, which is usually referred to as "AI winter."

Paul Werbos proposed that backpropagation can be used in neural networks [5]. He showed how to train multilayer perceptrons in his PhD thesis (1974), but due to the AI winter, it required a decade for researchers to work in this area. In 1986, this approach became popular with "Learning representations by back-propagating errors" by Rumelhart et al. [6]. First time in 1989, it was applied to a computer vision task which is handwritten digit classification [7]. It has demonstrated excellent performance on this task. However, after a short while, researchers started to face problems with the backpropagation algorithm. Deep (multilayer) neural networks trained with backpropagation did not work very well and particularly did not work as well as networks with fewer layers. It turned out that the magnitudes of

Figure 2.
Mark I Perceptron at the Cornell Aeronautical Laboratory, hardware implementation of the first perceptron (source: Cornell University Library [3]).

backpropagated errors shrink very rapidly and this prevents earlier layers to learn, which is today called as "the vanishing gradient problem." Again it took more than a decade for computers to handle more complex tasks. Some people prefer to name this period as the second AI winter.

Later, it was discovered that the initialization of weights has a critical importance for training, and with a better choice of nonlinear activation function, we can avoid the vanishing gradient problem. In the meantime, our computers got faster (especially thanks to GPUs), and huge amount of data became available for many tasks. G. Hinton and two of his graduate students demonstrated the effectiveness of deep networks at a challenging AI task: speech recognition. They managed to improve on a decade-old performance record on a standard speech recognition dataset. In 2012, a CNN (again G. Hinton and students) won against other machine learning approaches at the Large Scale Visual Recognition Challenge (ILSVRC) image classification task for the first time.

2.2 Working principles of a deep neural network

Technically any neural network with two or more hidden layers is "deep." However, in papers of recent years, deep networks correspond to the ones with many more layers. We show a simple network in **Figure 3**, where the first layer is the input layer, the last layer is the output layer, and the ones in between are the hidden layers.

In **Figure 3**, $a_j^{(i)}$ denotes the value after activation function is applied to the inputs in jth neuron of ith layer. If the predicted output of the network, which is $a_1^{(4)}$ in this example, is close to the actual output, then the "loss" is low. Previously mentioned backpropagation algorithm uses derivatives to carry the loss to the previous layers. $\frac{\partial L}{\partial a_1^{(4)}}$ represents the derivative of loss with respect to $a_1^{(4)}$, whereas $\frac{\partial L}{\partial a_1^{(2)}}$ represents the derivative of loss with respect to a second layer neuron $a_1^{(2)}$. The derivative of loss with respect to $a_1^{(2)}$ means how much of the final error (loss) is neuron $a_1^{(2)}$ responsible for.

Activation function is the element that gives a neural network its nonlinear representation capacity. Therefore, we always choose a nonlinear function. If activation function was chosen to be a linear function, each layer would perform a linear mapping of the input to the output. Thus, no matter how many layers were there, since linear functions are closed under composition, this would be equivalent to having a single (linear) layer.

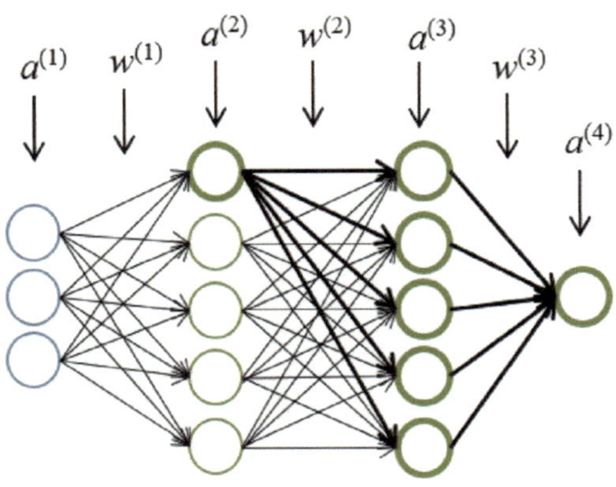

Figure 3.
A simple neural network with two hidden layers. Entities plotted with thicker lines are the ones included in Eq. (1), which will be used to explain the vanishing gradient problem.

The choice of activation function is critically important. In early days of multi-layer networks, people used to employ *sigmoid* or *tanh*, which cause the problem named as vanishing gradient. Let's explain the vanishing gradient problem with the network shown in **Figure 3**.

$$\frac{\partial L}{\partial {a_1}^{(2)}} = w^{(2)} \cdot \sigma'\left(z^{(3)}\right) \cdot w^{(3)} \cdot \sigma'\left({z_1}^{(4)}\right) \cdot \frac{\partial L}{\partial {a_1}^{(4)}} \tag{1}$$

Eq. (1) shows how the error in the final layer is backpropagated to a neuron in the first hidden layer, where $w^{(i)}$ denotes the weights in layer i and ${z_j}^{(i)}$ denotes the weighted input to the jth neuron in layer i. Here, let's assume *sigmoid* is used as the activation function. Then, ${a_j}^{(i)}$ denotes the value after the activation function is applied to ${z_j}^{(i)}$, i.e., ${a_j}^{(i)} = \sigma\left({z_j}^{(i)}\right)$. Finally, let σ' denote the derivative of *sigmoid* function. Entities in Eq. (1) are plotted with thicker lines in **Figure 3**.

Figure 4 shows the derivative of *sigmoid*, where we observe that the highest point derivative is equal to 25% of its original value. And most of the time, derivative is much less. Thus, at each layer $w^{(j)} \cdot \sigma'\left(z^{(j+1)}\right) \leq 0.25$ in Eq. (1). As a result, products decrease exponentially. $\frac{\partial L}{\partial {a_1}^{(2)}}$ becomes 16 (or more) times smaller than $\frac{\partial L}{\partial {a_1}^{(4)}}$. Thus, gradients become very small (vanish), and updates on weights get smaller, and they begin to "learn" very slowly. Detailed explanation of the vanishing gradient problem can be found in [8].

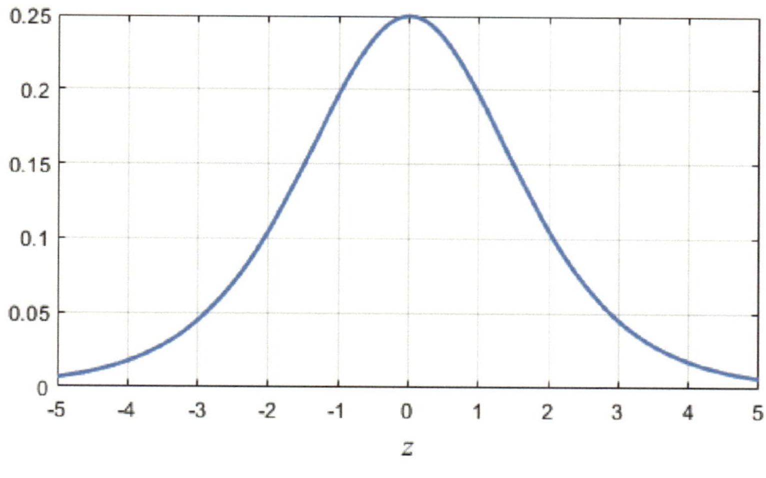

Figure 4.
Derivative of the sigmoid function.

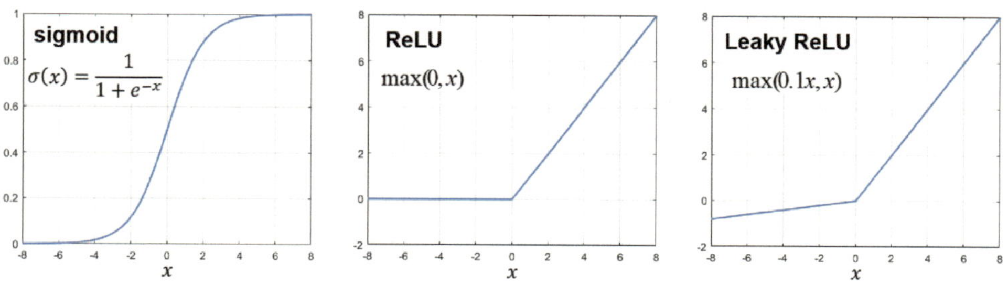

Figure 5.
Plots for some activation functions. Sigmoid is on the left, rectified linear unit is in the middle, and leaky rectified linear unit is on the right.

Today, choices of activation function are different. A rectified linear unit (ReLU), which outputs zero for negative inputs and identical value for positive inputs, is enough to eliminate the vanishing gradient problem. To gain some other advantages, leaky ReLU and parametric ReLU (negative side is multiplied by a coefficient) are among the popular choices (**Figure 5**).

3. Natural language processing

Deep learning transformed the field of natural language processing (NLP). This transformation can be described by better representation learning through newly proposed neural language models and novel neural network architectures that are fine-tuned with respect to an NLP task.

Deep learning paved the way for neural language models, and these models introduced a substantial performance improvement over n-gram language models. More importantly, neural language models are able to learn good representations in their hidden layers. These representations are shown to capture both semantic and syntactic regularities that are useful for various downstream tasks.

3.1 Representation learning

Representation learning through neural networks is based on the distributional hypothesis: "words with similar distributions have similar meanings" [9] where distribution means the neighborhood of a word, which is specified as a fixed-size surrounding window. Thus, the neighborhoods of words are fed into the neural network to learn representations implicitly.

Learned representations in hidden layers are termed as distributed representations [10]. Distributed representations are local in the sense that the set of activations to represent a concept is due to a subset of dimensions. For instance, cat and dog are hairy and animate. The set of activations to represent "being hairy" belongs to a specific subset of dimensions. In a similar way, a different subset of dimensions is responsible for the feature of "being animate." In the embeddings of both cat and dog, the local pattern of activations for "being hairy" and "being animate" is observed. In other words, the pattern of activations is local, and the conceptualization is global (e.g., cat and dog).

The idea of distributed representation was realized by [11] and other studies relied on it. Bengio et al. [11] proposed a neural language model that is based on a feed-forward neural network with a single hidden layer and optional direct connections between input and output layers.

The first breakthrough in representation learning was word2vec [12]. The authors removed the nonlinearity in the hidden layer in the proposed model architecture of [11]. This model update brought about a substantial improvement in computational complexity allowing the training using billions of words. Word2vec has two variants: continuous bag-of-words (CBOW) and Skip-gram.

In CBOW, a middle word is predicted given its context, the set of neighboring left and right words. When the input sentence "creativity is intelligence having fun" is processed, the system predicts the middle word "intelligence" given the left and right contexts (**Figure 6**). Every input word is in one-hot encoding where there is a vocabulary size (V) vector of all zeros except the one in that word's index. In the single hidden layer, instead of applying a nonlinear transformation, the average of the neighboring left and right vectors (w_c) is computed to represent the context. As the order of words is not taken into consideration by averaging, it is named as a bag-of-words model. Then the middle word's (w_t) probability given the context

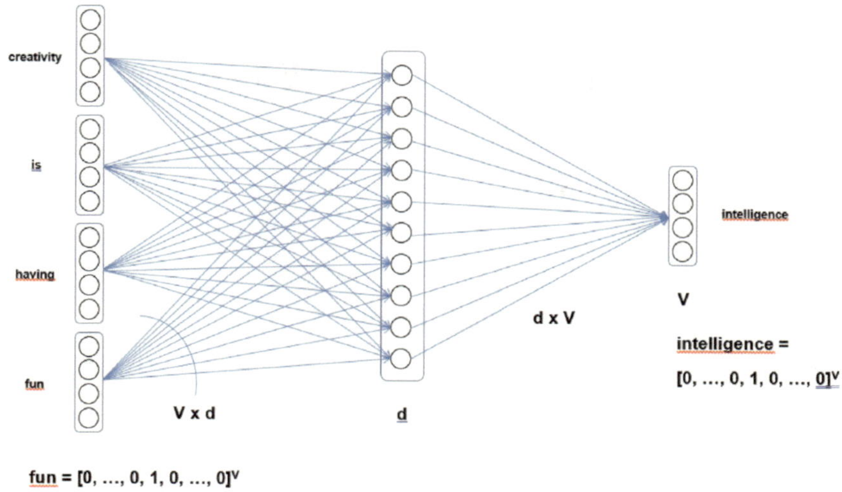

Figure 6.
CBOW architecture.

$(p(w_t|w_c))$ is calculated through softmax on context-middle word dot product vector (Eq. (2)). Finally, the output loss is calculated based on the cross-entropy loss between the system predicted output and the ground-truth middle word.

$$p(w_t|w_c) = \frac{exp\,(w_c \cdot w_t)}{\sum_{j \in V} exp\,(w_j \cdot w_t)} \tag{2}$$

In Skip-gram, the system predicts the most probable context words for a given input word. In terms of a language model, while CBOW predicts an individual word's probability, Skip-gram outputs the probabilities of a set of words, defined by a given context size. Due to high dimensionality in the output layer (all vocabulary words have to be considered), Skip-gram has higher computational complexity than CBOW (**Figure** 7). To deal with this issue, rather than traversing all vocabulary in the output layer, Skip-gram with negative sampling (SGNS) [13] formulates the problem as a binary classification where one class represents the current context's occurrence probability, whereas the other is all vocabulary terms' occurrence in the present context. In the latter probability calculation, a sampling approach is incorporated. As vocabulary terms are not distributed uniformly in contexts, sampling is performed from a distribution where the order of the frequency of vocabulary words in corpora is taken into consideration. SGNS incorporates this sampling idea by replacing the Skip-gram's objective function. The new objective function (Eq. (3)) depends on maximizing $P(D = 1|w, c)$, where w, c is the word-context pair. This probability denotes the probability of (w, c) coming from the corpus data. Additionally, $P(D = 0|u_i, c)$ should be maximized if (u_i, c) pair is not included in the corpus data. In this condition, (u_i, c) pair is sampled, as the name suggests negative sampled k times.

$$\sum_{w,c}\left(\log\sigma\left(\vec{w}\cdot\vec{c}\right)\right) + \sum_{i=1}^{k}\left(\log\sigma\left(\overrightarrow{-u_i}\cdot\vec{c}\right)\right) \tag{3}$$

Both word2vec variants produced word embeddings that can capture multiple degrees of similarity including both syntactic and semantic regularities.

A regular extension to word2vec model was doc2vec [14], where the main goal is to create a representation for different document levels, e.g., sentence and

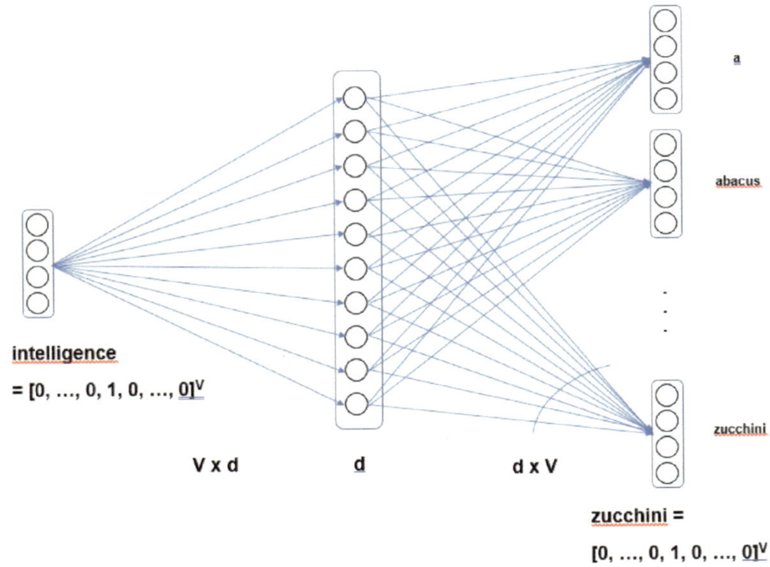

Figure 7.
Skip-gram architecture.

paragraph. Their architecture is quite similar to the word2vec except for the extension with a document vector. They generate a vector for each document and word. The system takes the document vector and its words' vectors as an input. Thus, the document vectors are adjusted with regard to all the words in this document. At the end, the system provides both document and word vectors. They propose two architectures that are known as distributed memory model of paragraph vectors (DM) and distributed bag-of-words model of paragraph vectors (DBOW).

DM: In this architecture, inputs are the words in a context except for the last word and document, and the output is the last word of the context. The word vectors and document vector are concatenated while they are fed into the system.

DBOW: The input of the architecture is a document vector. The model predicts the words randomly sampled from the document.

An important extension to word2vec and its variants is fastText [15], where they considered to use characters together with words to learn better representations for words. In fastText language model, the score between a context word and the middle word is computed based on all character n-grams of the word as well as the word itself. Here n-grams are contiguous sequences of n letters like unigram for a single letter, bigram for two consecutive letters, trigram for three letters in succession, etc. In Eq. (4), v_c represents a context vector, z_g is a vector associated with each n-gram, and G_w is the set of all character n-grams of the word w together with itself.

$$s(w,c) = \sum_{g \in G_w} \left(z_g^T v_c \right) \tag{4}$$

The idea of using the smallest syntactic units in the representation of words introduced an improvement in morphologically rich languages and is capable to compute a representation for out-of-vocabulary words.

The recent development in representation learning is the introduction of contextual representations. Early word embeddings have some problems. Although they can learn syntactic and semantic regularities, they are not so good in capturing a mixture of them. For example, they can capture the syntactic pattern *look-looks-looked*.

In a similar way, the words *hard, difficult*, and *tough* are embedded into closer points in the space. To address both syntactic and semantic features, Kim et al. [16] used a mixture of character- and word-level features. In their model, at the lowest level of hierarchy, character-level features are processed by a CNN; after transferring these features over a highway network, high-level features are learned by the use of a long short-term memory (LSTM). Thus, the resulting embeddings showed good syntactic and semantic patterns. For instance, the closest words to the word *richard* are returned as *eduard*, *gerard*, *edward*, and *carl*, where all of them are person names and have syntactic similarity to the query word. Due to character-aware processing, their models are able to produce good representations for out-of-vocabulary words.

The idea of capturing syntactic features at a low level of hierarchy and the semantic ones at higher levels was realized ultimately by the Embeddings from Language Models (ELMo) [17]. ELMo proposes a deep bidirectional language model to learn complex features. Once these features are learned, the pre-trained model is used as an external knowledge source to the fine-tuned model that is trained using task-specific data. Thus, in addition to static embeddings from the pre-trained model, contextual embeddings can be taken from the fine-tuned one.

Another drawback of previous word embeddings is they unite all the senses of a word into one representation. Thus, different contextual meanings cannot be addressed. The brand new ELMo and Bidirectional Encoder Representations from Transformers (BERT) [18] models resolve this issue by providing different representations for every occurrence of a word. BERT uses bidirectional Transformer language model integrated with a masked language model to provide a fine-tuned language model that is able to provide different representations with respect to different contexts.

3.2 NLP with neural network solutions

In NLP, different neural network solutions have been used in various downstream tasks.

Language data are temporal in nature so recurrent neural networks (RNNs) seem as a good fit to the task in general. RNNs have been used to learn long-range dependencies. However, because of the dependency to the previous time steps in computations, they have efficiency problems. Furthermore, when the length of sequences gets longer, an information loss occurs due to the vanishing gradient problem.

Long short-term memory architectures are proposed to tackle the problem of information loss in the case of long sequences. Gated recurrent units (GRUs) are another alternative to LSTMs. They use a gate mechanism to learn how much of the past information to preserve at the next time step and how much to erase.

Convolutional neural networks have been used to capture short-ranging dependencies like learning word representation over characters and sentence representation over its n-grams. Compared to RNNs, they are quite efficient due to independent processing of features. Moreover, through the use of different convolution filter sizes (overlapping localities) and then concatenation, their learning regions can be extended.

Machine translation is a core NLP task that has witnessed innovative neural network solutions that gained wide application afterwards. Neural machine translation aims to translate sequences from a source language into a target language using neural network architectures. Theoretically, it is a conditional language model where the next word is dependent on the previous set of words in the target sequence and the source sentence at the same time. In traditional language modeling, the next word's probability is computed based solely on the previous set of words. Thus, in conditional language modeling, conditional means conditioned on the source sequence's representation. In machine translation, source sequence's processing is termed as encoder part of the model, whereas the next word prediction task in the target language is called decoder. In probabilistic terms, machine translation aims to maximize the probability of the target sequence y given the source sequence x as follows.

$$\arg\max_{y} P(y|x) \tag{5}$$

This conditional probability calculation can be conducted by the product of component conditional probabilities at each time step where there is an assumption that the probabilities at each time step are independent from each other (Eq. (6)).

$$\begin{aligned} P(y|x) &= P(y_1|x)P(y_2|y_1,x)P(y_3|y_1,y_2,x), \dots, P(y_t|y_1, \dots, y_{t-1},x) \\ &= \prod_{i=1}^{t} P(y_i|y_1, \dots, y_{i-1},x) \end{aligned} \tag{6}$$

The first breakthrough neural machine translation model was an LSTM-based encoder-decoder solution [19]. In this model, source sentence is represented by the last hidden layer of encoder LSTM. In the decoder part, the next word prediction is based on both the encoder's source representation and the previous set of words in the target sequence. The model introduced a significant performance boost at the time of its release.

In neural machine translation, the problem of maximizing the probability of a target sequence given the source sequence can be broken down into two components by applying Bayes rule on Eq. (5): the probability of a source sequence given the target and the target sequence's probability (Eq. (7)).

$$\arg\max_{y} P(x|y)P(y) \tag{7}$$

In this alternative formulation, $P(x|y)$ is termed as translation model and $P(y)$ is a language model. Translation model aims to learn correspondences between source and target pairs using parallel training corpus. This learning objective is related to the task of learning word-level correspondences between sentence pairs. This alignment task is vital in that a correct translation requires to generate the counterpart word(s) for the local set of words in the source sentence. For instance, the French word group "tremblement de terre" must be translated into English as the word "earthquake," and these correspondences must be learned in the process.

Bandanau et al. [20] propose an attention mechanism to directly connect to each word in the encoder part in predicting the next word in each decoder step. This mechanism provides a solution to alignment in that every word in translation is predicted by considering all words in the source sentence, and the predicted word's correspondences are learned by the weights in the attention layer (**Figure 8**).

Attention is a weighted sum of values with respect to a query. The learned weights serve as the degree of query's interaction with the values at hand. In the case of translation, values are encoder hidden states, and query is decoder hidden state at the current time step. Thus, weights are expected to show each translation step's grounding on the encoder hidden states.

Eq. (8) gives the formulae for an attention mechanism. Here h_i represents each hidden state in the encoder (VALUES in **Figure 8**), w_i is the query vector coming

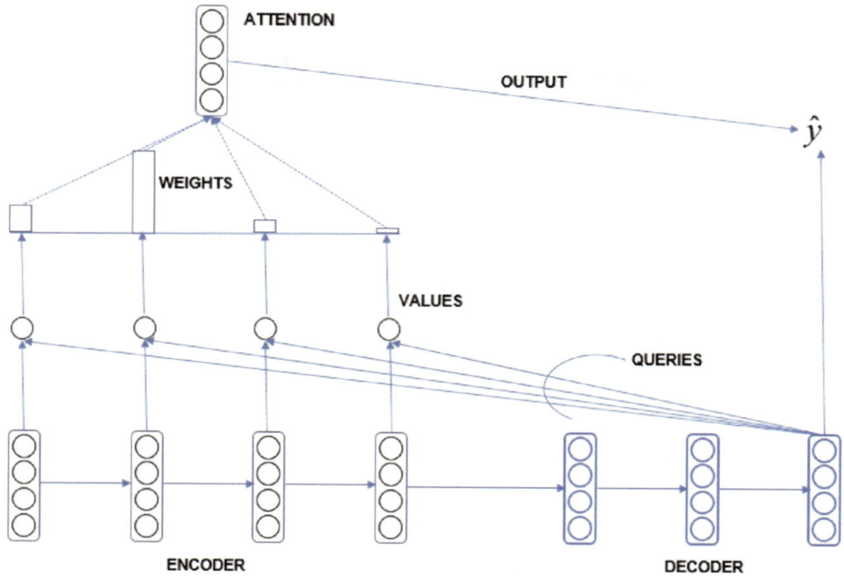

Figure 8.
Sequence-to-sequence attention.

from the current hidden state of the decoder (each QUERY in **Figure 8**), α_i (WEIGHTS in **Figure 8**) are attention weights, and K (OUTPUT in **Figure 8**) is the attention output that is combined with the last hidden state of the decoder to make the next word prediction in translation.

$$\begin{aligned} \alpha_i &= \frac{exp\,(h_i \cdot w_i)}{\sum_j exp\,(h_j \cdot w_j)} \\ o_i &= \alpha_i h_i \\ K &= \sum_i o_i \end{aligned} \tag{8}$$

The success of attention in addressing alignment in machine translation gave rise to the idea of a sole attention-based architecture called Transformer [21]. The Transformer architecture produced even better results in neural machine translation. More importantly, it has become state-of-the-art solution in language modeling and started to be used as a pre-trained language model. The use of it as a pre-trained language model and the transfer of this model's knowledge to other models introduced performance boost in a wide variety of NLP tasks.

The contribution of attention is not limited to the performance boost introduced but is also related to supporting explainability in deep learning. The visualization of attention provides a clue to the implicit features learned for the task at hand.

4. Computer vision and CNNs

To observe the performance of the developed methods on computer vision problems, several competitions are arranged all around the world. One of them is Large Scale Visual Recognition Challenge [22]. This event contains several tasks which are image classification, object detection, and object localization. In image classification task, the aim is to predict the class of images in the test set given a set of discrete labels, such as dog, cat, truck, plane, etc. This is not a trivial task since different images of the same class have quite different instances and varying viewpoints, illumination, deformation, occlusion, etc.

All competitors in ILSVRC train their model on ImageNet [22] dataset. ImageNet 2012 dataset contains 1.2 million images and 1000 classes. Classification performances of proposed methods were compared according to two different evaluation criteria which are top 1 and top 5 score. In top 5 criterion, for each image top 5 guesses of the algorithm are considered. If actual image category is one of these five labels, then the image is counted as correctly classified. Total number of incorrect answers in this sense is called top 5 error.

An outstanding performance was observed by a CNN (convolutional neural network) in 2012. AlexNet [23] got the first place in classification task achieving 16.4% error rate. There was a huge difference between the first (16.4%) and second place (26.1%). In ILSVRC 2014, GoogleNet [24] took the first place achieving 6.67% error rate. Positive effect of network depth was observed. One year later, ResNet took the first place achieving 3.6% error rate [25] with a CNN of 152 layers. In the following years, even lower error rates were achieved with several modifications. Please note that the human performance on the image classification task was reported to be 5.1% error [22].

4.1 Architecture of a typical CNN

CNNs are the fundamental structures while working on images and videos. A typical CNN is actually composed of several layers interleaved with each other.

4.1.1 Convolutional layer

Convolutional layer is the core building block of a CNN. It contains plenty of learnable filters (or kernels). Each filter is convolved across width and height of input images. At the end of training process, filters of network are able to identify specific types of appearances (or patterns). A mathematical example is given to illustrate how convolutional layers work (**Figure 9**). In this example, a 5×5 RGB image is given to the network. Since images are represented as 3D arrays of numbers, input consists of three matrices. It is convolved with a filter of size $3 \times 3 \times 3$ (height, weight, and depth). In this example, convolution is applied by moving the filter one pixel at a time, i.e., stride size = 1. First convolution operation can be seen at **Figure 9a**. After moving the kernel one pixel to the right, second convolution operation can be seen at **Figure 9b**. Element-wise multiplication $\odot$ is applied in each convolution phase. Thus the operation in **Figure 9a** is shown below Eq. (9).

$$
\begin{aligned}
&\begin{bmatrix} 1 & 1 & 0 \\ 1 & 1 & 2 \\ 2 & 1 & 1 \end{bmatrix} \odot \begin{bmatrix} 1 & 1 & 0 \\ 0 & 0 & 1 \\ -1 & 1 & 1 \end{bmatrix} + \begin{bmatrix} 1 & 2 & 1 \\ 1 & 0 & 0 \\ 2 & 2 & 0 \end{bmatrix} \odot \begin{bmatrix} 0 & 0 & 1 \\ 1 & 1 & 0 \\ 0 & -1 & -1 \end{bmatrix} \\
&\quad + \begin{bmatrix} 2 & 2 & 0 \\ 0 & 1 & 2 \\ 0 & 1 & 1 \end{bmatrix} \odot \begin{bmatrix} 1 & 1 & 0 \\ 1 & 1 & 0 \\ 0 & -1 & -1 \end{bmatrix} + 1 = 8
\end{aligned} \tag{9}
$$

Convolution depicted in **Figure 9** is performed with one filter which results in one matrix (called activation map) in the convolution layer. Using n filters for the input in **Figure 9** produces a convolution layer of depth n (**Figure 10**).

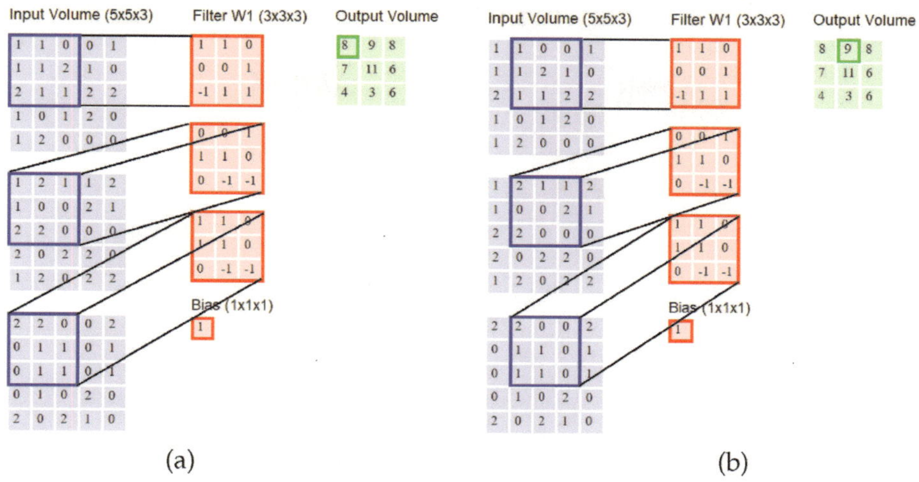

Figure 9.
Convolution process. (a) First convolution operation applied with filter W1. Computation gives us the top-left member of an activation map in the next layer. (b) Second convolution operation, again applied with filter W1.

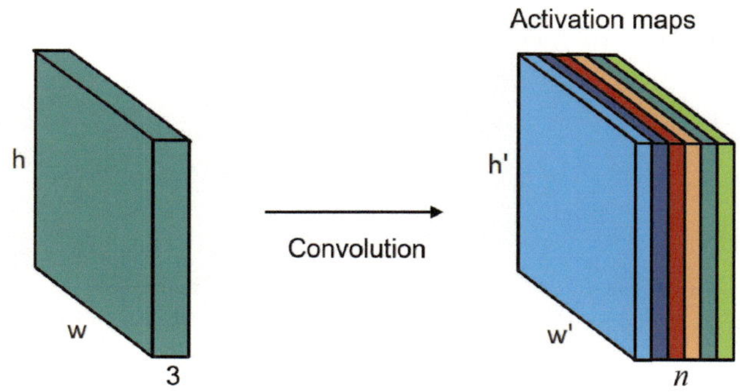

Figure 10.
Formation of a convolution layer by applying n *number of learnable filters on the previous layer. Each activation map is formed by convolving a different filter on the whole input. In this example input to the convolution is the RGB image itself (depth = 3). For every further layer, input is its previous layer. After convolution, width and height of the next layer may or may not decrease.*

4.1.2 Pooling layer

Pooling layer is commonly used between convolutional layers to reduce the number of parameters in the upcoming layers. It makes the representations smaller and the algorithm much faster. With max pooling, filter takes the largest number in the region covered by the matrix on which it is applied. Example input, on which 2×2 max pooling is applied, is shown in **Figure 11**. If the input size is $w \times h \times n$, then the output size is $(w/2) \times (h/2) \times n$. Techniques such as min pooling and average pooling can also be used.

Standard CNNs generally have several convolution layers, followed by pooling layers and at the end a few fully connected layers (**Figure 12**). CNNs are similar to standard neural networks, but instead of connecting weights to all units of the previous layer, a convolution operation is applied on the units (voxels) of the previous layer. It enables us scale weights in an efficient way since a filter has a fixed number of weights and it is independent of the number of the voxels in the previous layer.

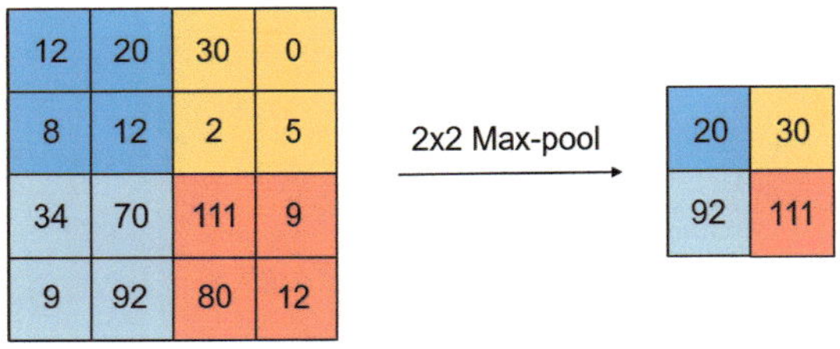

Figure 11.
Max pooling.

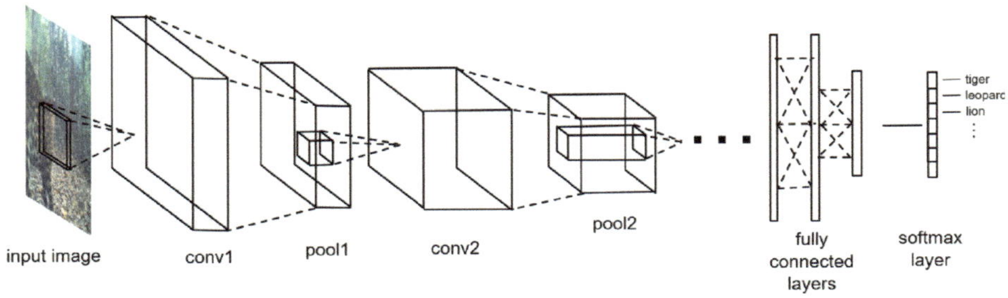

Figure 12.
A typical CNN for image classification task.

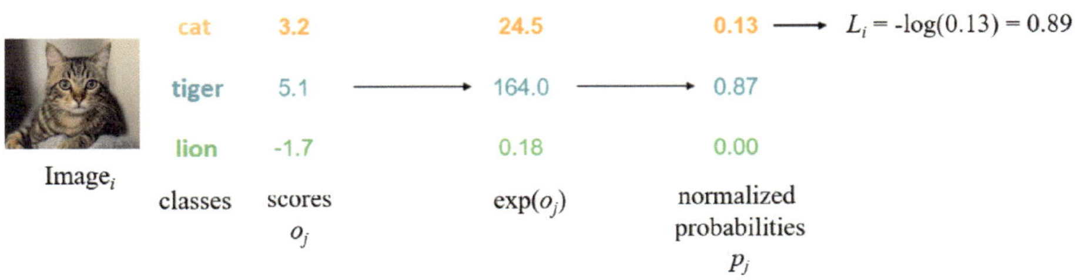

Figure 13.
An example of softmax classification loss calculation. Computed loss, L_i, is only for the ith sample in the dataset.

4.1.3 Classification layer

What we have in the last fully connected layer of a classification network is the output scores for each class. It may seem trivial to select the class with the highest score to make a decision; however we need to define a loss to be able to train the network. Loss is defined according to the scores obtained for the classes. A common practice is to use softmax function, which first converts the class scores into normalized probabilities (Eq. (10)):

$$p_j = \frac{e^{o_j}}{\sum_k e^{o_k}} \tag{10}$$

where k is the number of classes, o_j are the output neurons (scores), and p_j are the normalized probabilities. Softmax loss is equal to the log of the normalized probability of the correct class. An example calculation of softmax loss with three classes is given in **Figure 13**.

4.2 Generalization capability of CNNs

The ability of a model to make correct predictions for new samples after trained on the training set is defined as generalization. Thus, we would like to train a CNN with a high generalization capacity. Its high accuracy should not be only for training samples. In general, we should increase the size and variety of the training data, and we should avoid training an excessively complex model (simply called overfitting). Since it is not always easy to obtain more training data and to pick the best complexity for our model, let's discuss a few popular techniques to increase the generalization capacity.

4.2.1 Regularization loss

This is a term, $R(W)$, added to the data loss with a coefficient (λ) called regularization strength (Eq. (10)). Regularization loss can be a sum of L1 or L2 norm of weights. The interpretation of $R(W)$ is that we want smaller weights to be able to achieve smoother models for better generalization. It means that no input dimension can have a very large influence on the scores all by itself.

$$L = \frac{1}{N}\sum_{i=1}^{N} L_i + \lambda \cdot R(W) \tag{11}$$

4.2.2 Dropout

Another way to prevent overfitting is a technique called dropout, which corresponds to removing some units in the network [26]. The neurons which are "dropped out" in this way do not contribute to the forward pass (computation of loss for a given input) and do not participate in backpropagation (**Figure 14**). In each forward pass, a random set of neurons are dropped (with a hyperparameter of dropping probability, usually 0.5).

4.2.3 Data augmentation

The more training samples for a model, the more successful the model will be. However, it is rarely possible to obtain large-size datasets either because it is hard to collect more samples or it is expensive to annotate large number of samples. Therefore, to increase the size of existing raw data, producing synthetic data is sometimes preferred. For visual data, data size can be increased by rotating the picture at different angles, random translations, rotations, crops, flips, or altering brightness and contrast [27].

4.3 Transfer learning

Short after people realized that CNNs are very powerful nonlinear models for computer vision problems, they started to seek an insight of why these models

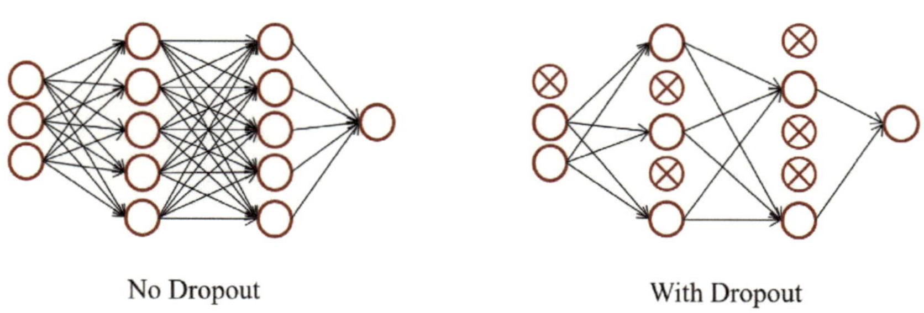

Figure 14.
Applying dropout in a neural net.

perform so well. To this aim, researchers proposed visualization techniques that provide an understanding of what features are learned in different layers of a CNN [28]. It turns out that first convolutional layers are responsible for learning low-level features (edges, lines, etc.), whereas as we go further in the convolutional layers, specific shapes and even distinctive patterns can be learned (**Figure 15**).

In early days of observing the great performance of CNNs, it was believed that one needs a very large dataset in order to use CNNs. Later, it was discovered that, since the pre-trained models already learned to distinguish some patterns, they provide great benefits for new problems and new datasets from varying domains. Transfer learning is the name of training a new model with transferring weights from a related model that had already been trained.

If the dataset in our new task is small but similar to the one that was used in pre-trained model, then it would work to change the classification layer (according to our classes) and train this last layer. However, if our dataset is also big enough, we can include a few more layers (starting from the fully connected layers at the end) to our retraining scheme, which is also called fine-tuning. For instance, if a face recognition model trained with a large database is available and you would like to use that model with the faces in your company, that would constitute an ideal case of transferring the weights from the pre-trained model and fine-tune one or two layers with your local database. On the other hand, if the dataset in our new task is not similar to the one used in pre-trained model, then we would need a larger dataset and need to retrain a larger number of layers. An example of this case is learning to classify CT (computer tomography) images using a CNN pre-trained on ImageNet dataset. In this situation, the complex patterns (cf. **Figure 15c** and **d**) that were learned within the pre-trained model are not much useful for your new task. If both the new dataset is small and images are much different from those of a trained model, then users should not expect any benefit from transferring weights. In such cases users should find a way to enlarge the dataset and train a CNN from scratch using the newly collected training data. The cases that a practitioner may encounter from the transfer learning point of view are summarized in **Table 1**.

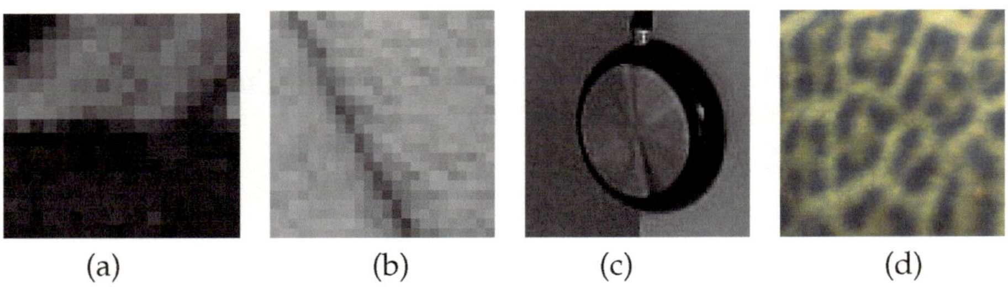

Figure 15.
Image patches corresponding to the highest activations in a random subset of feature maps. First layer's high activations occur at patches of distinct low-level features such as edges (a) and lines (b); further layers' neurons learn to fire at more complex structures such as geometric shapes (c) or patterns on an animal (d). Since activations in the first layer correspond to small areas on images, resolution of patches in (a) and (b) is low.

	Very similar dataset	Very different dataset
Very little data	Replace the classification layer	Not recommended
A lot of data	Fine-tune a few layers	Fine-tune a larger number of layers

Table 1.
Strategies of transfer learning according to the size of the new dataset and its similarity to the one used in pre-trained model.

Figure 16.
Example images for each class used in the experiment of transfer learning for animal classification.

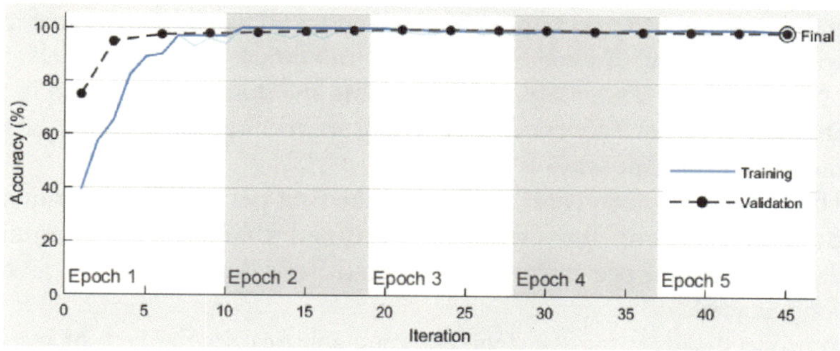

(a)

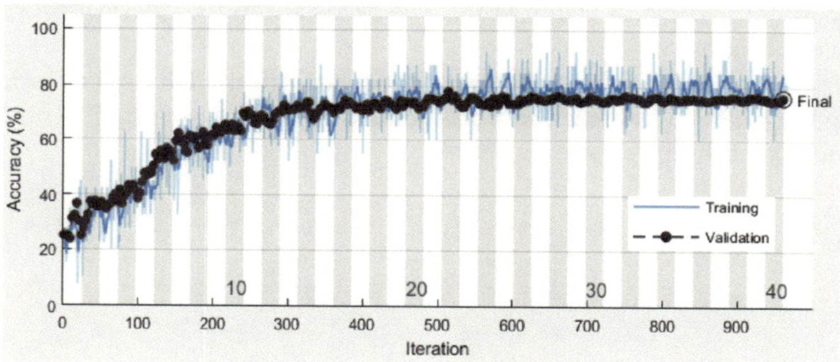

(b)

Figure 17.
Training and validation set accuracies obtained (a) with transfer learning and (b) without transfer learning.

To emphasize the importance of transfer learning, let us present a small experiment where the same model is trained with and without transfer learning. Our task is the classification of animals (four classes) from their images. Classes are zebra, leopard, elephant, and bear where each class has 350 images collected from the Internet (**Figure 16**). Transfer learning is performed using an AlexNet [23] pre-trained on ImageNet dataset. We have replaced the classification layer with a four-neuron layer (one for each class) which was originally 1000 (number of classes in ImageNet). In training conducted with transfer learning, we reached a 98.81% accuracy on the validation set after five epochs (means after seeing the dataset five times during training). Readers can observe that accuracy is quite satisfactory even after one epoch (**Figure 17a**). On the other hand, in training without transfer learning, we could reach only 76.90% accuracy even after 40 epochs (**Figure 17b**). Trying different hyperparameters (regularization strength, learning rate, etc.) could have a chance to increase accuracy a little bit more, but this does not alleviate the importance of applying transfer learning.

5. Conclusions

Deep learning has become the dominant machine learning approach due to the availability of vast amounts of data and improved computational resources. The main transformation was observed in text and image analysis.

In NLP, change can be described in two major lines. The first line is learning better representations through ever-improving neural language models. Currently, self-attention-based Transformer language model is state-of-the-art, and learned representations are capable to capture a mix of syntactic and semantic features and are context-dependent. The second line is related to neural network solutions in different NLP tasks. Although LSTMs proved useful in capturing long-term dependencies in the nature of temporal data, the recent trend has been to transfer the pre-trained language models' knowledge into fine-tuned task-specific models. Self-attention neural network mechanism has become the dominant scheme in pre-trained language models. This transfer learning solution outperformed existing approaches in a significant way.

In the field of computer vision, CNNs are the best performing solutions. There are very deep CNN architectures that are fine-tuned, thanks to huge amounts of training data. The use of pre-trained models in different vision tasks is a common methodology as well.

One common disadvantage of deep learning solutions is the lack of insights due to learning implicitly. Thus, attention mechanism together with visualization seems promising in both NLP and vision tasks. The fields are in the quest of more explainable solutions.

One final remark is on the rise of multimodal solutions. Till now question answering has been an intersection point. Future work are expected to be devoted to multimodal solutions.

Author details

Selma Tekir*† and Yalin Bastanlar†
Computer Engineering Department, Izmir Institute of Technology, Izmir, Turkey

*Address all correspondence to: selmatekir@iyte.edu.tr

† These authors are contributed equally.

References

[1] McCulloch WS, Pitts W. A logical calculus of the ideas immanent in nervous activity. Bulletin of Mathematical Biophysics. 1943;**5**: 115-133

[2] Rosenblatt F. The Perceptron, a Perceiving and Recognizing Automaton Project Para. Cornell Aeronautical Laboratory; 1957

[3] Images from the Rare Book and Manuscript Collections. Cornell University Library. Available from: https://digital.library.cornell.edu/catalog/ss:550351

[4] Minsky M, Papert S. Perceptrons. An Introduction to Computational Geometry. Cambridge, MA: MIT Press; 1969

[5] Werbos P. Beyond regression: New tools for prediction and analysis in the behavioral sciences [PhD thesis]. Cambridge, MA: Harvard University; 1974

[6] Rumelhart DE, Hinton GE, Williams RJ. Learning representations by back-propagating errors. Nature. 1986;**323**:533-536

[7] LeCun Y, Jackel LD, Boser B, Denker JS, Graf HP, Guyon I, et al. Handwritten digit recognition: Applications of neural network chips and automatic learning. IEEE Communications Magazine. 1989; **27**(11):41-46

[8] Nielsen M. Neural Network and Deep Learning. Available from: http://neuralnetworksanddeeplearning.com/chap5.html [Accessed: 30 December 2019]

[9] Harris Z. Distributional structure. Word. 1954;**10**(23):146-162

[10] Hinton GE, McClelland JL, Rumelhart DE. Distributed representations. In: Parallel Distributed Processing: Explorations in the Microstructure of Cognition. Vol. 1. Cambridge, MA: MIT Press; 1986. pp. 77-109

[11] Bengio Y, Ducharme R, Vincent P, Janvin C. A neural probabilistic language model. Journal of Machine Learning Research. 2003;**3**:1137-1155

[12] Mikolov T, Chen K, Corrado G, Dean J. Efficient estimation of word representations in vector space. In: Workshop Proceedings of ICLR. 2013

[13] Mikolov T, Sutskever I, Chen K, Corrado G, Dean J. Distributed representations of words and phrases and their compositionality. CoRR. abs/1310.4546. Available from: http://arxiv.org/abs/1310.4546

[14] Le Q, Mikolov T. Distributed representations of sentences and documents. In: Proceedings of the 31st International Conference on International Conference on Machine Learning (ICML). 2014

[15] Bojanowski P, Grave E, Joulin A, Mikolov T. Enriching word vectors with subword information. Transactions of the Association for Computational Linguistics (TACL). 2016;**5**:135-146

[16] Kim Y, Jernite Y, Sontag D, Rush AM. Character-aware neural language models. In: Proceedings of Thirtieth AAAI Conference on Artificial Intelligence (AAAI). 2016

[17] Peters ME, Neumann M, Iyyer M, Gardner M, Clark C, Lee K, et al. Deep contextualized word representations. In: Proceedings of NAACL. 2018

[18] Devlin J, Chang M, Lee K, Toutanova K. BERT: Pre-training of deep bidirectional transformers for

language understanding. In: Proceedings of NAACL. 2019

[19] Sutskever I, Vinyals O, Le QV. Sequence to sequence learning with neural networks. In: Proceedings of Advances in Neural Information Processing Systems (NIPS). 2014

[20] Bandanau D, Cho K, Bengio Y. Neural machine translation by jointly learning to align and translate. In: Proceedings of 3rd International Conference on Learning Representations (ICLR). 2015

[21] Vaswani A, Shazeer N, Parmar N, Uszkoreit J, Jones L, Gomez AN, et al. Attention is all you need. In: Proceedings of Advances in Neural Information Processing Systems (NIPS). 2017

[22] Russakovsky O et al. ImageNet large scale visual recognition challenge. International Journal of Computer Vision (IJCV). 2015;**115**(3):211-252

[23] Krizhevsky A, Sutskever I, Hinton GE. Imagenet classification with deep convolutional neural networks. In: Proceedings of NIPS. 2012

[24] Szegedy C, Liu W, Jia Y, Sermanet P, Reed S, Anguelov D, et al. Going deeper with convolutions. In: Proceedings of CVPR. 2015

[25] He K, Zhang X, Ren S, Sun J. Deep residual learning for image recognition. In: Proceedings of CVPR. 2016

[26] Srivastava N et al. Dropout: A simple way to prevent neural networks from overfitting. The Journal of Machine learning Research. 2014;**15**(1): 1929-1958

[27] Goodfellow I, Bengio Y, Courville A. Deep Learning. Cambridge, MA: MIT Press; 2016

[28] Zeiler MD, Fergus R. Visualizing and understanding convolutional networks. In: Proceedings of ECCV. 2014

Chapter 2

Contribution to Decision Tree Induction with Python: A Review

Bouchra Lamrini

Abstract

Among the learning algorithms, one of the most popular and easiest to understand is the decision tree induction. The popularity of this method is related to three nice characteristics: interpretability, efficiency, and flexibility. Decision tree can be used for both classification and regression kind of problem. Automatic learning of a decision tree is characterised by the fact that it uses logic and mathematics to generate rules instead of selecting them based on intuition and subjectivity. In this review, we present essential steps to understand the fundamental concepts and mathematics behind decision tree from training to building. We study criteria and pruning algorithms, which have been proposed to control complexity and optimize decision tree performance. A discussion around several works and tools will be exposed to analyze the techniques of variance reduction, which do not improve or change the representation bias of decision tree. We chose *Pima Indians Diabetes* dataset to cover essential questions to understand pruning process. The paper's original contribution is to provide an up-to-date overview that is fully focused on implemented algorithms to build and optimize decision trees. This contributes to evolve future developments of decision tree induction.

Keywords: decision tree, induction learning, classification, pruning, bias-variance trade-off

1. Introduction

Decision tree induction is the most known and developed model of machine learning methods often used in data mining and business intelligence for prediction and diagnostic tasks [1–4]. It is used in classification problems, regression problems or time-dependent prediction. The main strength of decision tree induction is its interpretability characteristics. It is a graphical method designed for problems involving a sequence of decisions and successive events. More precisely, his results formalise the reasoning that an expert could have to reproduce the sequence of decisions and find a characteristic of an object. The main advantage of this type of model is that a human being can easily understand and reproduce decision sequence to predict the target category of a new instance. The results provide a graphic structure or a base of rules facilitates understanding and corresponds to human reasoning.

Learning by decision tree is part of supervised learning, where the class of each object in the database is given. The goal is to build a model from a set of examples associated with the classes to find a description for each of the classes from the common properties between the examples. Once this model has been built, we can extract a set of classification rules. In this model, the extracted rules are then used to classify new objects whose class is unknown. The classification is done by travelling a path from the root to a leaf. The class returned (default class) is the one that is most frequent among the examples on the sheet. At each internal node (decision node) of the tree, there is a test (question) which corresponds to an attribute in the learning base and a branch corresponding to each of the possible values of the attribute. At each leaf node, there is a class value. A path from the root to a node therefore corresponds to a series of attributes (questions) with their values (answers). This flowchart-like structure with recursive partitioning helps user in decision-making. It is this visualisation, which easily mimics the human-level thinking. That is why decision trees are easy to understand and interpret.

Another advantage of decision tree induction is its ability to automatically identify the most discriminating features for an use case, i.e., the most representative data inputs for a given task. This is explained by its flexibility and autonomy as a model with little assumption on the hypothesis space. It is an approach that remains particularly useful for input space problems and a powerful tool able to handle very large-scale problems, thus particularly useful in big data mining. However, it is generally less accurate than other machine learning models like neural networks.

In brief, this learning algorithm has the following three essential characteristics:

- Interpretability: Because of its flowchart-like structure, the way attributes interact to give a prediction is very readable.

- Efficiency: The induction process is done by a top-down algorithm which recursively splits terminal nodes of the current tree until they all contain elements of only one class. Practically, the algorithm is very fast in terms of running time and can be used on very large datasets (e.g. of millions of objects and thousands of features).

- Flexibility: This method does not make any hypothesis about the problem under consideration. It can handle both continuous and discrete attributes. Predictions at leaf nodes may be symbolic or numerical (in which case, trees are called regression trees). In addition, the tree induction method can be easily extended by improving tests at tree nodes (e.g. introducing linear combinations of attributes) or providing a prediction at terminal nodes by means of another model.

The review is organised into three parts. The first aims at introducing a brief history of decision tree induction. We present mathematically basics and search strategy used to train and build a decision tree. We discuss the supervised learning problem and the trade-off between a model's ability to minimise bias and variance. In this regard, we are extending our investigation to fundamental aspects, such as ensemble meta-algorithms and pruning methods, which we must put in advance for building an optimal decision tree. In the second section, we introduce some results obtained by means of the *Scikit-Learn Python* modules and *Pima Indians Diabetes* data in order to feed our discussions and our perspectives in terms of future developments and applications of Python community. The third section is devoted to the improvements of decision tree induction in order to improve its performance. We have collected some technical discussions that we raise given our experience in Research and Development (R&D). Finally, the conclusions give a general synthesis of the survey developed and discuss some ideas for future works.

2. A brief history of decision tree induction

There are many induction systems that build decision trees. Hunt et al. [5] were the first in this field to study machine learning using examples. Their concept learning system (CLS) framework builds a decision tree that tries to minimise the cost of classifying an object. There are two types of costs: (1) the cost of determining the value of a property of the object O_A exhibited by the object and (2) the misclassification cost of deciding that the object belongs to class C when its real class is K. The CLS method uses a strategy called *Lookahead* which consists of exploring the space of all possible decision trees to a fixed depth and choosing an action to minimise the cost in this limited space and then moving one level down in the tree. Depending on the depth of the *Lookahead* chosen, CLS can require a substantial amount of computation but has been able to unearth subtle patterns in the objects shown to it.

Quinlan [6] proposed Iterative Dichotomiser 3 (ID3), which takes up certain concepts of CLS. ID3 was developed following a challenge induction task on the study of end of chess games posed by Donald Michie. Analogue concept learning system (ACLS) [7] is a generalisation of ID3. CLS and ID3 require that each attribute used to describe the object takes its values in a finite set. In addition to this type of attribute, ACLS allows the use of attributes whose values can be integer. ASSISTANT [8], which is a descendant of ID3, allows the use of continuous attributes and builds a binary decision tree. ASSISTANT avoids overfitting by using a pruning technique, which has resulted in ASSISTANT-86 [9]. Another descendant of ID3 is [10, 11], which will be explained later.

There is another family of induction systems, such as the algorithm of the star AQ [12], which induces a set of decision rules from a base of examples. AQ builds an R function that covers positive examples and rejects negative ones. CN2 [13] learns a set of unordered rules of the form "IF-THEN" from a set of examples. For this, CN2 performs a top-down search (from general to specific) in the rule space, looking for the best rule, then removes the examples covered by this rule and repeats this process until no good rule is found. CN2's strategy is similar to that of AQ in that it eliminates the examples covered by the discovered rule, but it also differs in that it specialises a starting rule instead of generalising it.

Statisticians have attributed the authorship of decision tree building to Morgan and Sonquist [1], who are the first researchers to introduce the automatic interaction detector (AID) method. This method is applied to learning problems whose attribute to predict (the class) is quantitative. It works sequentially and is independent of the extent of linearity in the classifications or the order in which the explanatory factors are introduced. Morgan and Sonquist were among the first to use decision trees and among the first to use regression trees.

Several extensions have been proposed: theta AID (THAID) [2] and chi-squared AID (CHAID) [3] which uses chi-square as the independence gap to choose the best partitioning attribute. There is also a method proposed by [4] called classification and regression tree (CART) which builds a binary decision tree using the feature and threshold that yield the largest information gain at each node.

Quinlan [11] then proposes the *C4.5* algorithm for IT community. C4.5 removed the restriction that entities must be categorical by dynamically defining a discrete attribute based on numerical variables. This discretization process splits the continuous attribute value into a discrete set of intervals. C4.5 then converts the trees generated at the end of learning step into sets of if-then rules. This accuracy of each rule is well taken into account to determine the order in which they must be applied. Pruning is performed by removing the rule's precondition if the precision of the rule improves without it.

Many decision tree algorithms have been developed over the years, for example, SPRINT by Shafer et al. [14] and SLIQ by Mehta et al. [15]. One of the studies comparing decision trees and other learning algorithms was carried out by Tjen-Sien et al. [16]. The study shows that C4.5 has a very good combination of error rate and speed. C4.5 assumes that the training data is in memory. Gehrke et al. [17] proposed Rainforest, an approach to develop a fast and scalable algorithm. In [18], Kotsiantis represents a synthesis of the main basic problems of decision trees and current research work. The references cited cover the main theoretical problems that can lead the researcher into interesting directions of research and suggest possible combinations of biases to explore.

2.1 Mathematical basics and search strategy

The automatic learning of the rules in a decision tree consists in separating the learning objects into disjoint sub-samples of objects (which have no elements in common) where the majority of objects ideally have the same value for the output variable, i.e. the same class in the case of a classification problem. Each internal node performs a test on an input attribute. This test is determined automatically based on the initial training sample and according to test selection procedures that differ from one tree induction algorithm to another. For attributes with numerical values (or after encoding data), the test consists in comparing the value of an attribute with a numerical value which is called discretization threshold. According to the algorithm used, the terminal nodes of the tree are labelled either by the majority class of objects in the training sample which have reached this sheet following successive separations or by a distribution of probabilities of the classes by frequency of these objects in each class.

As indicated above, the main learning algorithms using decision trees are C4.5 [11] and CART [4]. The CART algorithm is very similar to C4.5, except for a few properties [19, 20], but it differs in that it supports numerical target variables (regression) and does not compute rule sets. The CART algorithm can be used to construct classification and regression decision trees, depending on whether the dependent variable is categorical or numeric. It also handles missing attribute values. The decision tree built by the CART algorithm is always a binary decision tree (each node has only two child nodes), also called hierarchical optimal discriminate analysis (HODA). The measurement of impurity (or purity) used in the decision tree by CART is the Gini index (C4.5 uses the notion of entropy) for classification tasks. In regression tasks, the fit method takes inputs and target arguments as in the classification setting, only that in this case target, it is expected to have floating point values (continuous values) instead of integer values. For a leaf L_i, common criteria to minimise as for determining locations for future splits are mean squared error (MSE), which minimises the $L_i + 1$ error using mean values at terminal nodes, and mean absolute error (MAE), which minimises the L_i error using median values at terminal nodes.

Several software for decision trees building are available, most of them referenced in the literature. We cite the chi-squared automatic interaction detector (CHAID) method implemented in the SIPINA[1] tool which seeks to produce a tree of limited size, allowing to initiate a data exploration. WEKA[2] uses C4.5 algorithm, and there is no need to discretize any of the attributes, and scikit-learn Python library uses an optimised version of the CART algorithm. The current (version

[1] http://eric.univ-lyon2.fr/~ricco/sipina.html

[2] https://www.cs.waikato.ac.nz/ml/weka/

0.22.1) implementation of scikit-learn library does not support categorical variables. A data encoding is mandatory at this stage (the labels transform into a value between 0 and nbClasses-1). The algorithm options are described in the Python documentation[3].

The algorithm below generally summarises the learning phase of a decision tree which begins at the top of the tree with a root node containing all the objects of the learning set:

Algorithm 1: _build_DT

if *DT contains objects all of which belong to the same class* **then**
  return a leaf labeled with this class
else
  1- Take $[a_k < a_{th}]$ = _choose_test_(DT);
  2- Split DT into DT_{left} and DT_{right} according to test $[a_k < a_{th}]$ and build the sub-trees SDT_{left} = build DT_{left} and SDT_{right} = build DT_{right} from this sub-sets;
  3- Creat a node with the test $[a_k < a_{th}]$, make SDT_{left} and SDT_{right} like successors of this node and return the resulting tree.
end
_choose_test_(DT): Select an attribute a_k and a threshold a_{th} which minimizes the measurement of the score on DT.

In order for the tree to be easily interpreted, its size must be minimum. Thus, the test selection procedure applied at each node aims to choose the test (the attribute-threshold pair) which separates the objects from the current sub-sample in an optimal way, i.e. which reduces the uncertainty linked to the output variable within successor nodes. An entropy measurement (score based on a normalisation of the Shannon information measurement) allows to evaluate the gain of information provided by the test carried out. Once the model has been built, we can infer the class of a new object by propagating it in the tree from top to bottom according to the tests performed. The chosen test separates the current sample of objects into two sub-samples which are found in the successors of this node. Each test at a node makes it possible to direct any object to one of the two successors of this node according to the value of the attribute tested at this node. In other words, a decision tree is seen as a function which attributes to any object the class associated with the terminal node to which the object is directed following tests to the internal nodes of the tree. **Figure 1** illustrates an example using two input attributes with the partitioning of the input space it implies.

The induction algorithm continues to develop the tree until the terminal nodes contain sub-samples of objects that have the same output value. The label associated with a leaf in the tree is determined from the objects in the learning set that have been directed to this leaf. The majority class among the classes of these objects can be used or even the distribution of class probabilities if a stop criterion has interrupted development before reaching "pure" nodes.

The principal objective of an induction algorithm is to build on the learning data a simpler tree whose reliability is maximum, i.e. the classification error rate is minimal. However, a successful and very precise model on the learning set is not necessarily generalizable to unknown examples (objects), especially in the presence of noisy data. In this case, two sources of error expressed in the form of bias (difference between the real value and the estimated value) and the variance can generally influence the precision of a model. Several bibliographic analyses ([21]

---

[3] https://scikit-learn.org/stable/modules/tree.html#

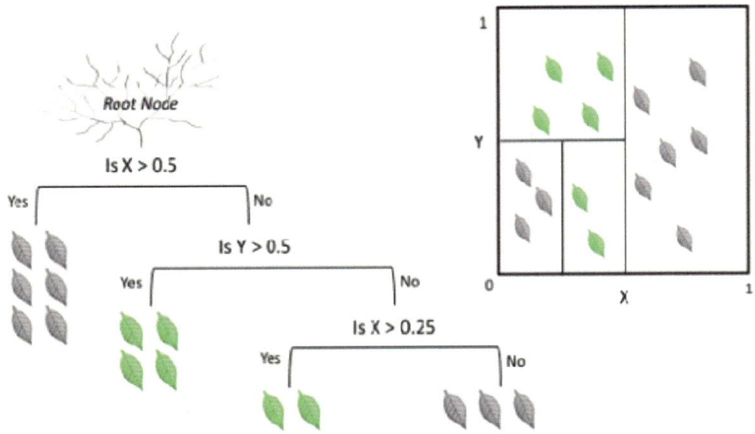

Figure 1.
An example of a decision tree and the partition it implies (Figure taken from https://www.kdnuggets.com/ website).

and the references cited in this work, [22]) have shown that decision trees suffer from a significant variance which penalises the precision of this technique. A tree may be too large due to too many test nodes determined at the bottom of the tree on sub-samples of statistically unreliable size objects. The choice of tests (attributes and thresholds) at the internal nodes of a decision tree can also depend on a sample to another which contributes to the variance of the models built. For these reasons, the criteria for stopping the development of a tree or simplification techniques such as pruning procedures is to find a good compromise between the complexity of the model and its reliability on an independent sample. These techniques can only improve the first source of error (bias) mentioned above. Different variance reduction techniques are proposed in the literature, notably the ensemble meta-algorithms such as bagging, random forests, extra-trees and boosting.

The ensemble meta-algorithms are effective in combination with decision trees. These methods differ by their way of adapting the original tree induction algorithm and/or aggregating the results. Bagging, random forests and extra-trees methods have several similarities. They independently build T constitutive trees. The predictions of different trees are aggregated as follows: each tree produces a vector of class probabilities. The T trees probability are additional in a weight vector, and the class that receives the most weight according to this one is assigned to the object. Note that these three methods use a random component and their precision can then vary slightly from one execution to another. Boosting method produces sequentially (and deterministically) the set of trees unlike these three methods using a different aggregation procedure. These methods have been successfully applied to numerous applications, notably in bioinformatics [23] and in networks [24]. Maree [22] presents a bibliographical analysis of these methods. His work covers the problem of automatic image classification using sets of random trees combined with a random extraction of sub-windows of pixel values.

2.2 Pruning

Pruning is a model selection procedure, where the models are the pruned sub-trees of the maximum tree T_0. Let T be the set of all binary sub-trees of T having the same root as T_0. This procedure minimises a penalised criterion where the penalty is proportional to the number of leaves in the tree [25]. Defining the optimal size of a decision tree consists in stopping the *pre-pruning* or reducing the *post-pruning* of the tree to have the best pruned sub-tree from the maximum tree to the sense of the generalisation error, i.e. improving the predictive aspect of the tree, on the one hand, and reducing its complexity, on the other hand. To this end, several pruning methods have been developed, such as:

- *Minimal cost complexity pruning* (*MCCP*), also called as *post-pruning* for the CART algorithm [4]. This method consists in constructing a nested sequence of sub-trees using a formulation called minimum cost complexity. In Section 2.2.1, we detail the general concept of this method that Scikit-Learn Library adopted in its implementation.

- *Reduced error pruning* (*REP*) consists of estimating the real error of a given sub-tree on a pruning or test set. The pruning algorithm is performed as follows: "As long as there is a tree that can be replaced by a leaf without increasing the estimate of the real error, then prune this tree". This technique gives a slightly congruent tree in the sense that some examples may be misclassified. The study of Elomaa and Kääriäinen [26] presents a detailed analysis of the REP method. In this analysis, the two authors evoke that the REP method was introduced by Quinlan [27] but the latter never presented it in an algorithmic way, which is a source of confusion. Even though REP is considered a very simple, almost trivial algorithm for pruning, many different algorithms have the same name. There is no consensus whether *REP* is a bottom-up algorithm or an iterative method. Moreover, it is not apparent that the training or pruning set is used to determine the labels of the leaves that result from pruning.

- *Pessimistic error pruning* (*PEP*). In order to overcome the disadvantages of the previous method, Quinlan [27] proposed a pruning strategy which uses a single set of construction and pruning of the tree. The tree is pruned by examining the error rate at each node and assuming that the true error rate is considerably worse. If a given node contains N records in which E among them are misclassified, then the error rate is estimated at E/N. The central concern of the algorithm is to minimise this estimate, by considering this error rate as a very optimistic version of the real error rate [28, 29].

- Minimum error pruning (MEP) was proposed by Niblett and Bratko [30], critical value pruning (CVP) by Mingers [31] and error-based pruning (EBP) proposed by Quinlan as an improvement of the PEP method, for the algorithm C4.5.

2.2.1 Pre-pruning

Pre-pruning consists in fixing a stopping rule which allows to stop the growth of a tree during learning phase by fixing a local stopping criterion which makes it possible to evaluate the informational contribution of the segmentation relating to the node that is being processed. The principle of the CHAID algorithm [32] is based on the same principle by accepting segmentation if the measure of information gain (χ^2 difference in independence or t from Tschuprow [3]) calculated on a node is significantly higher than a chosen threshold. According to Rakotomalala et al. [32, 33], formalisation involves a test of statistical hypothesis: the null hypothesis is the independence of the segmentation variable with the class attribute. If the calculated χ^2 is higher than the theoretical threshold corresponding to the risk of the first kind that we have set (respectively if the p-value calculated is lower than the risk of first kind), we accept the segmentation.

One of the cons of this algorithm is that it prematurely stops the building process of the tree. Furthermore, the use of the statistical test is considered critical. This is a classic independence test whose variable tested is produced at the end of several optimisation stages: search for the optimal discretization point for continuous variables and then search for the segmentation variable which maximises the measure used. The statistical law is no longer the same from one stage to another. The correction of the test by the introduction of certain procedures known as the *Bonferroni* correction [33] is recommended, but in practice, this type of correction does not lead to improvement in terms of classification performance. We also cite the work of [34], which proposes two pruning approaches: the first is a method of simplifying rules by the test of statistical independence to modify the pruning mechanism of the algorithm CHAID, and the second uses validation criteria inspired by the discovery technique of association rules.

The depth (maximum number of levels) of the tree and the minimum number of observations from which no further segmentation attempts are made also remain two practical options that can be fixed at start learning to manage the complexity of the model. However, the choice of these parameters remains a critical step in the tree building process because the final result depends on these parameters that we have chosen. To this is added the fact that the evaluation is local (limited to a node) and we take more account of the global evaluation of the tree. It is therefore necessary to propose a rule which is not too restrictive (respectively not too permissive) to obtain a suitable tree and not undersized (respectively not oversized).

2.2.2 Post-pruning

The algorithm for building a binary decision tree using CART browses for each node the m attributes $(x_1, x_2, \dots, x_m)$ one by one, starting with x_1 and continuing up to x_m. For each attribute, it explores all the possible tests (splits), and it chooses the best split (dichotomy) which maximises the reduction in impurity. Then, it compares the m best splits to choose the best of them. The function that measures impurity should reach its maximum when the instances are fairly distributed between the different classes and its minimum when a class contains all the examples (the node is pure). There are different functions which satisfy these properties. The function used by CART algorithm is Gini function (Gini impurity index). Gini function on a node t with a distribution of class probabilities on this node P (j|t), c = 1, ..., k is:

$$\begin{aligned} G(p) &= \phi(P(1|t), P(2|t), \dots, P(k|t)) \\ &= \sum_c P(c|t).(1 - P(c|t)) \end{aligned} \tag{1}$$

If a split s on a node t splits the subset associated with this node into two subsets, left t_G with a proportion p_G and right t_D with a proportion p_D, we can define the impurity reduction measure as follows:

$$\Delta G(s,t) = G(t) - p_G * G(t_G) - p_D * G(t_D) \tag{2}$$

On each node, if the set of candidate splits is S, the algorithm searches the best split s^* such that:

$$\Delta G(s^*, t) = max_{s \in S} \Delta G(s,t) \tag{3}$$

Suppose we got some *splits* and came up with a set of terminal nodes $\tilde{T}$. The set of *splits*, used in the same order, determines the binary tree T. We have $I(t) = G(t)p(t)$. So the impurity function on the tree is:

$$I(T) = \sum_{t \in \tilde{T}} I(t) = \sum_{t \in \tilde{T}} G(t) * p(t) \quad (4)$$

$G(t)$ is the impurity measure on the node t, and $p(t)$ is the probability that an instance belongs to the node t.

It is easy to see that the selection of the splits which maximise $\Delta_i(s,t)$ is equivalent to the selection of the splits which minimise the impurity $I(T)$ on all the trees. If we take any node $t \in \tilde{T}$ and we use a split s which partitions the node into two parts t_D and t_G, the new tree $\acute{T}$ has the following impurity:

$$I\left(\acute{T}\right) = \sum_{\tilde{T}-\{t\}} I(t) + I(t_D) + I(t_G) \quad (5)$$

Because we have partitioned the subset arrived at t and t_D and t_G, reducing the impurity of the tree is therefore:

$$I(T) - I\left(\acute{T}\right) = I(T) - I(t_D) - I(t_G) \quad (6)$$

It only depends on the node t and the *splits* s. So, to maximise the reduction of impurity in the tree on a node t, we maximise:

$$\Delta I(s,t) = I(t) - I(t_G) - I(t_D) \quad (7)$$

The proportions p_D are defined as follows: $p_D = p(t_D)/p(t)$, $p_G = p(t_G)/p(t)$ and $p_G + p_D = 1$. So, Eq. (7) can be written as follows:

$$\begin{aligned} \Delta I(s,t) &= \left[G(t) - p_G * G(t_G) - p_D * G(t_D)\right] * p(t) \\ &= \Delta G(s,t) * p(t) \end{aligned} \quad (8)$$

Since $p(t)$ is the only difference between (s,t) and (s,t), the same *splits* s^* maximises both expressions.

The stop splitting criterion used by CART was very simple: for a threshold $\beta > 0$, a node is declared terminal (leaf) if $max\Delta I(s,t) \leq \beta$. The algorithm assigns to each terminal node the most probable class.

Post-pruning is a procedure that appeared with the CART method [4]. It was very widely taken up in different forms thereafter. The principle is to build the tree in two phases. (1) The first phase of expansion consists in producing the purest possible trees and in which all segmentations are accepted even if they are not relevant. This is the principle of hurdling building. (2) In the second phase, we try to reduce the tree by using another criterion to compare trees of different sizes. The building time of the tree is of course longer. It can be penalising when the database is very large while the objective is to obtain a tree that performs better in classification phase.

The idea that was introduced by Breiman et al. [4] is to construct a sequence of trees $T_0, .., T_i, .., T_t$, which minimise a function called *cost complexity metric* (previously mentioned). This function combines two factors: the classification error rate and the number of leaves in the tree using α parameter.

For each internal node, *Ne* and T_0, the relationship is defined as:

$$\alpha(p) = \frac{\Delta R^S_{emp}}{|T_p| - 1} \tag{9}$$

where ΔR^S_{emp} is the number of additional errors that the decision tree makes on the set of samples S when we prune it at position p. $|p| - 1$ measures the number of sheets deleted. The tree T_{i+1} is obtained by pruning T_i at its node which has the smallest value of $\alpha(p)$ parameter. We thus obtain a sequence $T_0, .., T_i, .., T_t$ of elements of T, the last of which is reduced to a leaf. To estimate the error rate for each tree, the authors suggest using two different methods, one based on cross-validation and the other on a new test base.

3. Decision tree classifier building in Scikit-Learn

Today there are several best machine learning websites that propose tutorials to show how decision trees work using the different modules of python. We quote for example three popular websites: Towards Data Science,[4] KDnuggets,[5] and Kaggle.[6] Developers offer in a few lines of optimised code how to use decision tree method by covering the various topics concerning attribute selection measures, information gain, how to optimise decision tree performance, etc.

From our side, we choose *Pima Indians Diabetes* datasets (often used in classification problems) to examine the various tuned parameters proposed as arguments by Scikit-Learn package. The Pima are a group of Native Americans living in Arizona. A genetic predisposition allowed this group to survive normally to a diet poor of carbohydrates for years. In the recent years, a sudden shift from traditional agricultural crops to processed foods, together with a decline in physical activity, made them develop the highest prevalence of type 2 diabetes, and for this reason they have been subject of many studies. The original dataset is available at UCI Machine Learning Repository and can be downloaded from this address,[7] “diabetes-data.tar.Z”, containing the distribution for 70 sets of data recorded on diabetes patients, several weeks to months worth of glucose, insulin and lifestyle data per patient. The dataset includes data from 768 women with 8 characteristics, in particular:

1. Number of times pregnant (*NTP*)

2. Plasma glucose concentration in 2 h in an oral glucose tolerance test (*PGC*)

3. Diastolic blood pressure (mm Hg) (*DBP*)

4. Triceps skinfold thickness (mm) (*TSFT*)

5. Two-hour serum insulin (mu U/ml) (*HSI*)

6. Body mass index (weight in kg/(height in m)^2) (*BMI*)

---

[4] https://towardsdatascience.com/decision-tree-algorithm-explained-83beb6e78ef4

[5] https://www.kdnuggets.com/2020/01/decision-tree-algorithm-explained.html

[6] https://www.kaggle.com/dmilla/introduction-to-decision-trees-titanic-dataset

[7] http://archive.ics.uci.edu/ml/datasets/Pima+Indians+Diabetes

7. Diabetes pedigree function (*DPF*)

8. Age (years)

The last column of the dataset indicates if the person has been diagnosed with diabetes or not.

Without any data preparation step (cleaning, missing values processing, etc.), we partitioned the dataset into a training data (75%) to build the tree and test data (25%) for prediction. Then we kept the default settings which we can see through the profile class function (**Figure 2**).

The Scikit-Learn documentation[8] explains in detail how to use each parameter and offers other modules and functions to search information and internal structures of classifier from training to building step. Among these parameters, we highlight in this review the following four we use to optimise the tree:

- *criterion*: Optional (default = "gini"). This parameter allows to measure the quality of a split, use the different-different attribute selection measure and supports two criteria, "gini" for the Gini index and "entropy" for the information gain.

- max_depth: Optional (default = None), the maximum depth of a tree. If None, then nodes are expanded until all the leaves contain less than min_samples_split samples. A higher value of maximum depth causes overfitting, and a lower value causes underfitting.

- min_samples_leaf: Optional (default = 1), the minimum number of samples required to be at a leaf node. A split point at any depth will only be considered if it leaves at least min_samples_leaf training samples in each of the left and right branches. This may have the effect of smoothing the model, especially in regression.

- min_impurity_decrease: Optional (default = 0.0). A node will be split if this split induces a decrease of the impurity greater than or equal to this value. The weighted impurity decrease equation is the following:

$$N_t/N * (impurity - N_{tR}/N_t * right_impurity - N_{tL}/N_t * left_impurity) \quad (10)$$

where N is the total number of samples, N_t is the number of samples at the current node, N_{tL} is the number of samples in the left child and N_{tR} is the number of samples in the right child. N, N_t, N_{tR} and N_{tL} all refer to the weighted sum, if sample_weight is passed.

```
DecisionTreeClassifier(class_weight=None, criterion='gini', max_depth=None,
                       max_features=None, max_leaf_nodes=None,
                       min_impurity_decrease=0.0, min_impurity_split=None,
                       min_samples_leaf=1, min_samples_split=2,
                       min_weight_fraction_leaf=0.0, presort=False,
                       random_state=None, splitter='best')
```

Figure 2.
Default setting to create decision tree classifier without pruning.

[8] https://scikit-learn.org/stable/modules/generated/sklearn.tree.DecisionTreeClassifier.html

In this example, each internal node has a decision rule that divides the data. The node impurity is set by default at Gini ratio. A node is pure when all the objects belong to the same class, i.e. impurity = 0. The unpruned tree resulting from this setting is inexplicable and difficult to understand. In **Figures 3** and **4**, we will show you how to adjust some tuned parameters to get an optimal tree by pruning.

"export_graphviz" and "pydotplus" modules convert the decision tree classifier to a "dot" file and in "png/pdf/.." format. Using various options of this modules, you can adjust leaf colours and edit leaf content, important descriptors, etc. Personally, I really enjoyed doing it during my R&D works.

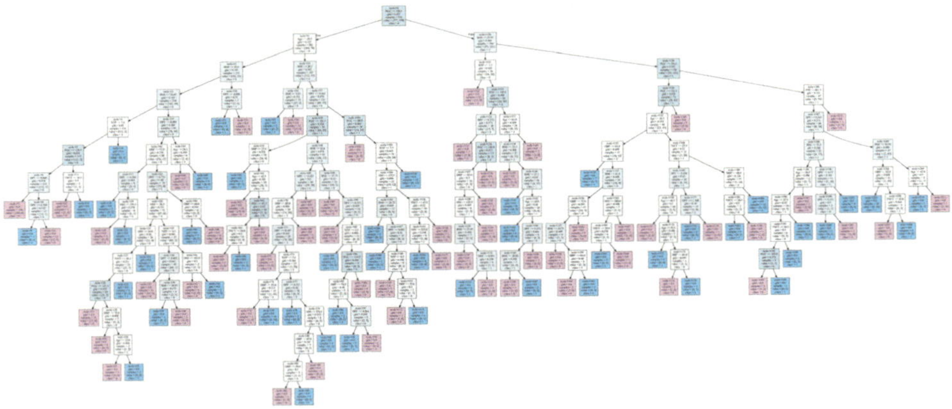

Figure 3.
Decision tree without pruning. Accuracy = 0.72.

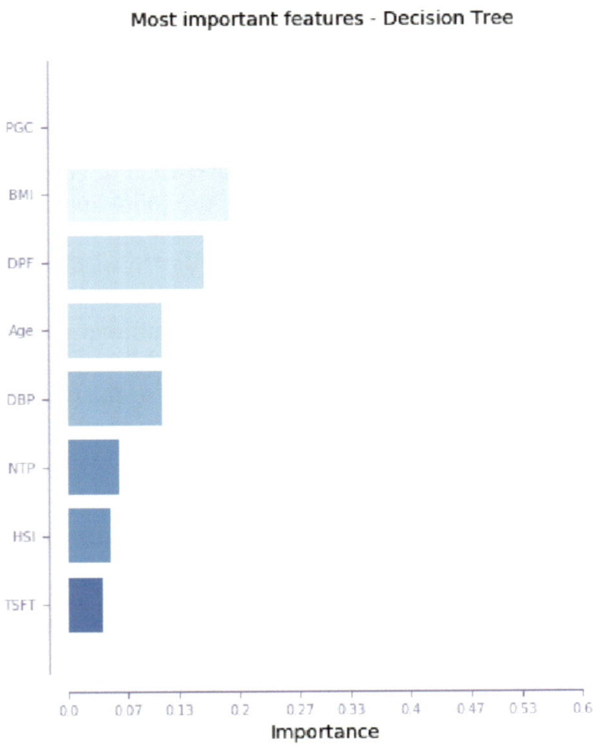

Figure 4.
Feature importance. Decision tree without pruning.

We will now adjust only one parameter, the maximum depth of the tree. This will control the tree size (number of levels). On the same data, we set maximum_depth at 4. Next, we set "min_impurity_decrease" at 0.01 and min_samples_leaf at 5. We will see that this pruned tree is less complex and easier to understand by a field expert than the previous flowchart. We will see that we have good accuracy with this setting. Accuracy can be computed by comparing actual test set values and predicted values (**Figures 5–8**).

Most important features of Pima Indians Diabetes dataset is shown in **Figures 4**, **6** and **8**. We can see the root node is glucose, which can show the glucose has the max information gain, so it confirm the common sense and the clinical

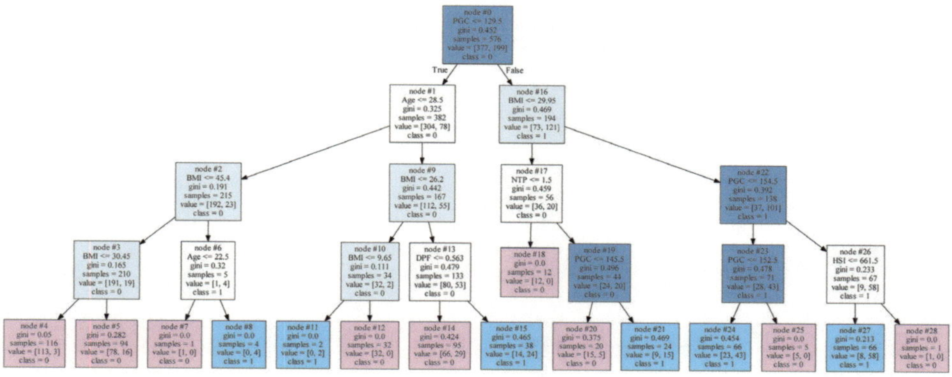

Figure 5.
Decision tree pruned by mean of maximum depth parameter. Accuracy = 0.79.

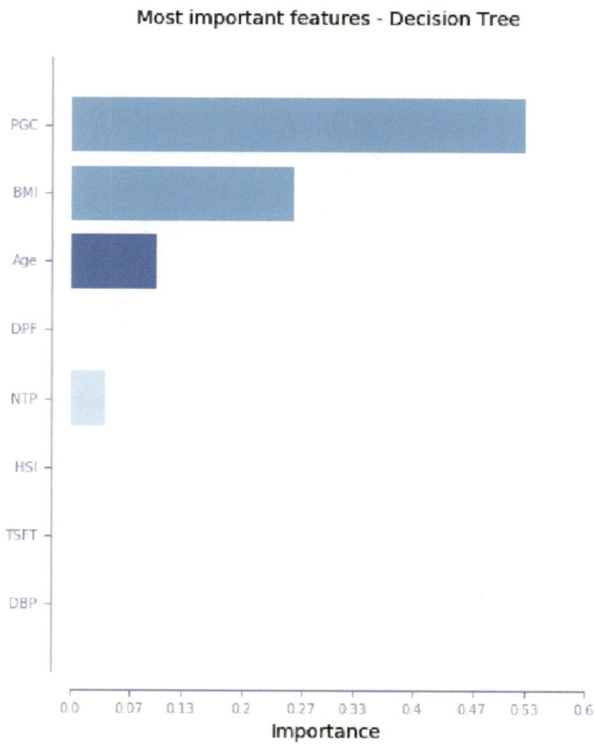

Figure 6.
*Feature importance. Decision tree after pruning (corresponding to **Figure** 5 results).*

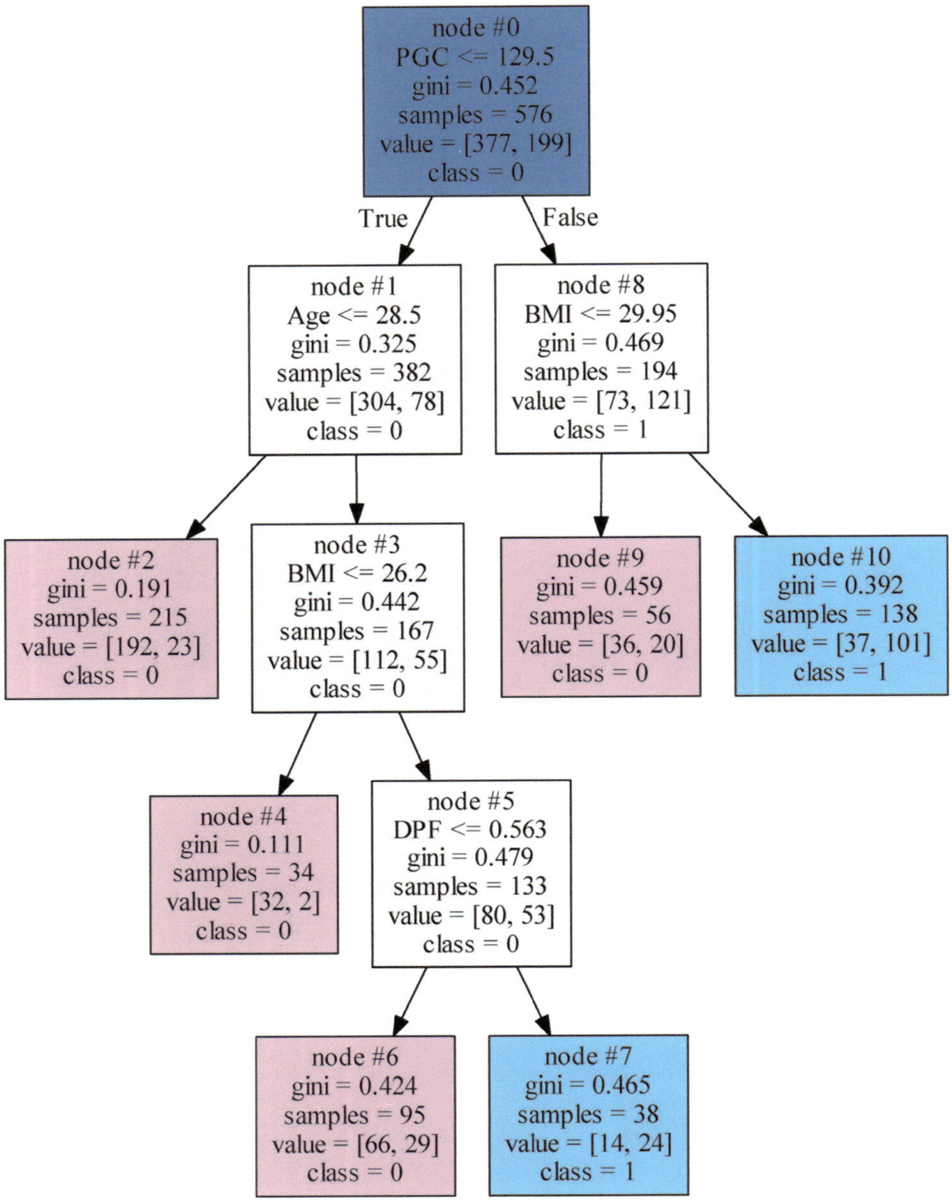

Figure 7.
Decision tree pruned by mean maximum depth and impurity parameters. Accuracy = 0.80.

diagnosis basis. Body mass index (BMI) and age are also found among the first important variables. According to consulting relevant information, we know there are three indicators to determine the diabetes mellitus, which are fasting blood glucose, random blood glucose and blood glucose tolerance. Pima Indians Diabetes dataset only has blood glucose tolerance. Prevalence of diabetes mellitus, hypertension and dyslipidaemia increase with higher BMI (BMI 25 kg/m^2). On the other hand, type 2 diabetes usually begins after age 40 and is diagnosed at an average age of 65. This is why the French National Authority for Health recommends renewing the screening test every 3 years in people over 45 years and every year if there is more than one risk factor.

Despite the stop criterion of tree depth, the trees generated may be too deep for a good practical interpretation. The notion of "accuracy" associated with each level

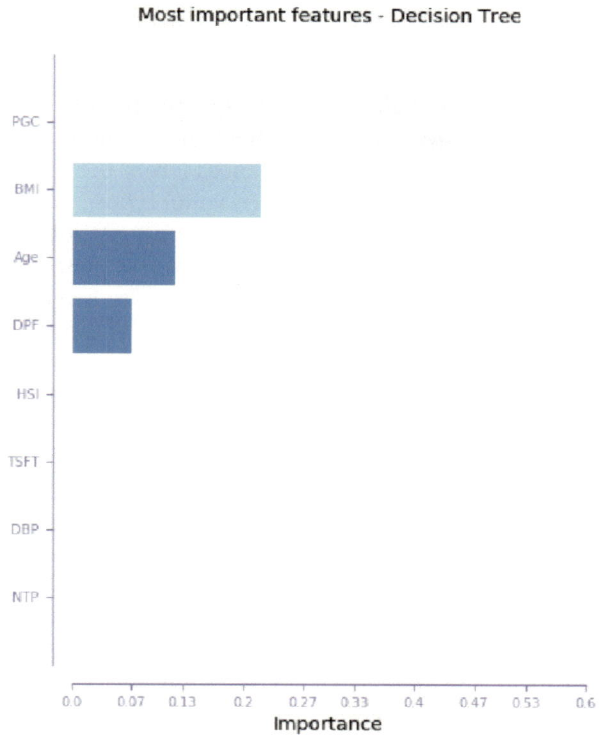

Figure 8.
*Feature importance. Decision tree after pruning (corresponding to **Figure 7** results).*

Prédit/réel	**Classe A**	**Classe B**
Classe A	VA (Vrais A)	FA (Faux A)
Classe B	FB (Faux B)	VB (Vrais B)

Table 1.
Confusion matrix.

of the tree will make it possible to present a partial tree sufficiently precise. This is based on the confusion matrix: The accuracy P is the ratio of well-classified elements to the sum of all elements and is defined by the following expression (**Table 1**):

$$\mathcal{P} = \frac{VA + VB}{VA + VB + FA + FB} \tag{11}$$

The accuracy associated with a level of the tree is calculated by summing the *VA*, *VB*, *FA* and *FB* taking into account the labels *A* or *B* of each node, and we add to *VA* or *VB* the elements corresponding to pure nodes *A* or *B* in the previous levels. We can thus decide to choose the partial tree according to the desired accuracy.

4. Discussions

Decision trees accept, like most learning methods, several hyper-parameters that control its behaviour. In our use case, we used Gini index like information criteria to split the learning data. This criterion has directed the method to build a tree with a maximum of 15 levels and to accept a node as a leaf if it includes at least five learning instances. Impurity (entropy) is a measure of disorder in dataset; if we have zero entropy, it means that all the instances of the target classes are the same, while it reaches its maximum when there is an equal number of instances of each class. At each node, we have a number of instances (from the dataset), and we measure its entropy. Setting impurity to a given value (chosen according to expertise and tests) will allow us to select the questions which give more homogeneous partitions (with the lowest entropy), when we consider only the instances for which the answer to the question is yes or no, that is to say when the entropy after answer to the question decreases.

During my previous R&D work, we used the CART algorithm implemented in the scikit-learn library. This implementation is close to the original one proposed by [4]; however there is no parameter for penalising the deviance of the model by its complexity (number of leaves) in order to build a sequence of trees nested in the prospect of optimal pruning by cross-validation. The generic function of k-fold cross-validation "GridSearchCV" can be used to optimise the depth parameter but with great precision in pruning. The depth parameter eliminates a whole level and not the only unnecessary leaves to the quality of the prediction. On the other hand, the implementation anticipates those of model aggregation methods by integrating the parameters (number of variables drawn, importance, etc.) which are specific to them. On the other hand, the graphical representation of tree is not included and requires the implementation of another free software like "Graphviz" and "Pydotplus" modules.

The pros and cons of decision trees are known and described in almost all the articles and works developed in this field. We highlight some that we consider important for industrial applications. Selecting features is an extremely important step when creating a machine learning solution. If the algorithm does not have good input functionality, it will not have enough material to learn, and the results will not be good, even if we have the best machine learning algorithm ever designed. The selection of characteristics can be done manually depending on the knowledge of the field and the machine learning method that we plan to use or by using automatic tools to evaluate and select the most promising. Another common problem with datasets is the problem of missing values. In most cases, we take a classic imputation approach using the most common value in the training data, or the median value. When we replace missing values, we should understand that we are modifying the original problem and be careful when using this data for other analytical purposes. This is a general rule in machine learning. When we change the data, we should have a clear idea of what we are changing, to avoid distorting the final results. Fortunately, decision tree requires fewer data preprocessing from users. It is used with missing data, and there is no need to normalise features. However, we must be careful in the way we describe the categorical data. Having a priori knowledge of the data field, we can favour one or more modalities of a descriptor to force the discretization process to choose a threshold, which highlights the importance of the variables. Moreover, Geurts [21] has shown that the choice of tests (attributes and thresholds) at the internal nodes of the tree can strongly depend on samples, which also contributes to the variance of the models built according to this method.

Decision tree can easily capture nonlinear patterns, which is important in big data processing. Nevertheless it is sensitive to noisy data, and it can overfit it. In big data mining, online data processing is subject to continuous development (upgrade, environment change, catching up, bugs, etc.) impacting the results expected by customers and users. To this, the problem of variation that can be reduced by bagging and boosting algorithms (that we mentioned in Section 2.1) is added.

Decision tree is biased with imbalance dataset. It is recommended to balance dataset before training to prevent the tree from being biased towards the classes that are dominant. According to scikit-learn documentation "class balancing can be done by sampling an equal number of samples from each class, or preferably by normalising the sum of the sample weights (sample_weight) for each class to the same value. Also note that weight-based pre-pruning criteria, such as min_weight_fraction_leaf, will then be less biased towards dominant classes than criteria that are not aware of the sample weights, like min_samples_leaf".

5. Conclusions

Decision trees simply respond to a classification problem. Decision tree is one of the few methods that can be presented quickly, without getting lost in mathematical formulations difficult to grasp, to hearing not specialised in data processing or machine learning. In this chapter, we have described the key elements necessary to build a decision tree from a dataset as well as the pruning methods, pre-pruning and post-pruning. We have also pointed to ensemble meta-algorithms as alternative for solving the variance problem. We have seen that letting the decision tree grow to the end causes several problems, such as overfitting. In addition, the deeper the tree is, the more the number of instances (samples) per leaf decreases. On the other hand, several studies have shown that pruning decreases the performance of the decision tree in estimating probability.

Decision tree properties are now well known. It is mainly positioned as a reference method despite the fact that efforts to develop the method are less numerous today. The references cited in this chapter are quite interesting and significant. They provide a broad overview of statistical and machine learning methods by producing a more technical description pointing the essential key points of tree building. In spite of the fact that the CART algorithm has been around for a long time, it remains an essential reference, by its precision, its exhaustiveness and the hindsight which the authors, developers and researchers demonstrate in the solutions they recommend. Academic articles also suggest new learning techniques and often use it in their comparisons to locate their work, but the preferred method in machine learning also remains C4.5 method. The availability of source code on the web justifies this success. C4.5 is now used for Coronavirus Disease 2019 (COVID-19) diagnosis [35, 36].

Finally, we would like to emphasise that the interpretability of a decision tree is a factor which can be subjective and whose importance also depends on the problem. A tree that does not have many leafs can be considered easily interpretable by a human. Some applications require good interpretability, which is not the case for all prediction applications. On industrial problems, an interpretable model with great precision is often necessary to increase knowledge of the field studied and identify new patterns that can provide solutions to needs and to several expert questions. We continue to put a lot of effort (scientific researchers, developers, experts, manufacturers, etc.) to make more improvements to this approach: decision tree induction. This chapter opens several opportunities in terms of algorithms and in terms of applications. For our use case, we would like to have more data to predict the type of diabetes and determine the proportion of each indicator, which can improve the accuracy of predicting diabetes.

Author details

Bouchra Lamrini
Ex-researcher (Senior R&D Engineer) within Livingobjects Company, Toulouse, France

*Address all correspondence to: lamrini.bouchra@gmail.com

References

[1] Morgan J, Sonquist J. Problems in the analysis of survey data, and a proposal. Journal of the American Statistical Association. 1963;**58**(2):415-435

[2] Morgan J, Messenger R. THAID-A Sequential Analysis Program for the Analysis of Nominal Scale Dependent Variables. Ann Arbor: Survey Research Center, Institute for Social Research, University of Michigan; 1973

[3] Kass G. An exploratory technique for investigating large quantities of categorical data. Applied Statistics. 1973; **29**(2):119-127

[4] Breiman L, Friedman J, Stone C, Olshen R. Classification and Regression Trees. Taylor & Francis;; 1984. Available from: https://books.google.fr/books?id=JwQx-WOmSyQC

[5] Hunt E, Marin J, Stone P. Experiments in Induction. New York, NY, USA: Academic Press; 1997. Available from: http://www.univ-tebessa.dz/fichiers/mosta/544f77fe0cf29473161c8f87.pdf

[6] Quinlan JR. Discovering rules by induction from large collections of examples. In: Michie D, editor. Expert Systems in the Micro Electronic Age. Vol. 1. Edinburgh University Press; 1979. pp. 168-201

[7] Paterson A, Niblett T. ACLS Manual. Rapport Technique. Edinburgh: Intelligent Terminals, Ltd; 1982

[8] Kononenko I, Bratko I, Roskar E. Experiments in Automatic Learning of Medical Diagnostic Rules. Technical Report. Ljubljana, Yugoslavia: Jozef Stefan Institute; 1984

[9] Cestnik B, Knononenko I, Bratko I. Assistant86-A knowledge elicitation tool for sophisticated users. In: Bratko I, Lavrac N, editors. Progress in Machine Learning. Wilmslow, UK: Sigma Press; 1987. pp. 31-45

[10] Quinlan JR, Compton PJ, Horn KA, Lazarus L. Inductive knowledge acquisition: A case study. In: Proceedings of the Second Australian Conference on Applications of Expert Systems. Boston, MA, USA: Addison Wesley Longman Publishing Co., Inc.; 1987. pp. 137-156

[11] Quinlan R. C4.5-Programs for Machine Learning. San Francisco, CA, USA: Morgan Kaufman; 1993. p. 368

[12] Michalski RS. On the quasi minimal solution of the general covering problem. In: Proceedings of the 5th International Symposium on Information Processing; 1969. pp. 125-128

[13] Clark P, Niblett T. The cn2 induction algorithm. Machine Learning. 1989;**3**(4):261283

[14] Shafer J, Agrawal R, Mehta M. SPRINT-A scalable parallel classifier for data mining. Proceeding of the VLDB Conference. 1996;**39**:261-283. DOI: 10.1007/s10462-011-9272-4

[15] Mehta M, Agrawal R, Riassnen J. SLIQ-A fast scalable classifier for data mining. Extending Database Technology. 1996;**39**:18-32

[16] Tjen-Sien L, Wei-Yin L, Yu-Shan S. SLIQ. A comparison of prediction accuracy, complexity, and training time of thirty-three old and new classification algorithms. Machine Learning. 2000;**40**:203-228

[17] Gehrke J, Ramakrishnan R, Ganti V. RainForest—A framework for fast decision tree construction of large datasets. Data Mining and Knowledge Discovery. 2000;**4**(2–3):127-162. DOI: 10.1023/A:1009839829793

[18] Kotsiantis SB. Decision trees—A recent overview. Artificial Intelligence Review. 2013;**39**:261-283. DOI: 10.1007/s10462-011-9272-4

[19] Rokach L, Maimon O. Top-down induction of decision trees classifiers—A survey. IEEE Transactions on Systems, Man, and Cybernetics: Part C. 2005;**35**(4):476-487. DOI: 10.1109/TSMCC.2004.843247

[20] Brijain MR, Patel R, Kushik MR, Rana K. International Journal of Engineering Development and Research. 2014;**2**(1):1-5. DOI: 10.1109/TSMCC.2004.843247

[21] Geurts P. Contributions to Decision Tree Induction: Bias/Variance Tradeoff and Time Series Classification. Belgium: University of Liège; 2002. p. 260. Available from: http://www.montefiore.ulg.ac.be/services/stochastic/pubs/2002/Geu02

[22] Marée R, Geurts P, Visimberga G, Piater J, Wehenkel L. A comparison of generic machine learning algorithms for image classification. In: Proceeding Research and Development in Intelligent Systems XX; 2004. pp. 169-182

[23] Geurts P, Fillet M, de Seny D, Meuwis MA, Merville MP, Wehenkel L. Proteomic mass spectra classification using decision tree based ensemble methods. Bioinformatics. 2004;**21**(14):3138-3145. DOI: 10.1093/bioinformatics/bti494

[24] Geurts P, Khayat E, Leduc G. A machine Learning approach to improve congestion control over wireless computer networks. In: Proceedings of the IEEE International Conference on Data Mining (ICDM'04); 2004. Available from: https://ieeexplore.ieee.org/document/1410316

[25] Genuer R, Poggi JM. Arbres CART et Forêts aléatoires, Importance et sélection de variables. hal-01387654v2; 2017. pp. 1-5. Available from: https://hal.archives-ouvertes.fr/hal-013876

[26] Elomaa T, Kääriäinen M. An analysis of reduced error pruning. Journal of Artificial Intelligence Research. 2001:163-187. Available from: http://dl.acm.org/citation.cfm?id=26079.26082

[27] Quinlan R. Simplifying decision trees. International Journal of Human Computer Studies. 1999;**51**(2):497-510

[28] Berry MJ, Linoff G. Data Mining Techniques for Marketing, Sales, and Customer Relationship Management. John Wiley & Sons; 1997. Available from: http://hdl.handle.net/2027/mdp.39015071883859

[29] Brostaux Y. Étude du Classement par Forêts Aléatoires D'échantillons Perturbés à Forte Structure D'interaction. Belgium: Faculté Universitaire des Sciences Agronomiques de Gembloux; 2005. p. 168. Available from: http://www.montefiore.ulg.ac.be/services/stochastic/pubs/2002/Geu02

[30] Niblett T, Bratko I. Learning decision rules in noisy domains. In: Proceedings of Expert Systems '86, The 6th Annual Technical Conference on Research and Development in Expert Systems III. Cambridge University Press; 1987. pp. 25-34. Available from: http://dl.acm.org/citation.cfm?id=26079.26082

[31] Mingers J. Experts systems-rule induction with statistical data. Journal of the Operational Research Society. 1987;**38**(1):39-47

[32] Rakotomalala R, Lallich S. Construction d'arbres de décision par optimisation. Revue des Sciences et Technologies de l'Information - Série RIA: Revue d'Intelligence Artificielle. 2002;**16**(6):685-703. Available from: https://hal.archives-ouvertes.fr/hal-00624091

[33] Rakotomalala R. Arbres de décision. Revue Modular. 2005;**33**:163-187. Available from: https://www.rocq.inria.fr/axis/modulad/archives/numero-33/tutorial-rakotomalala-33/rakotomalala-33-tutorial.pdf

[34] Mededjel M, Belbachir H. Post-élagage indirect des arbres de décision dans le data mining. In: 4th International Conference: Sciences of Electronic, Technologies of Information and Telecommunications; 2007. pp. 1-7. Available from: http://www.univ-tebessa.dz/fichiers/mosta/544f77fe0cf29473161c8f87.pdf

[35] Wiguna W, Riana D. Diagnosis of Coronavirus Disease 2019 (COVID-19) surveillance using C4.5 Algorithm. Jurnal Pilar Nusa Mandiri. 2020;**16**(1):71-80. DOI: 10.33480/pilar.v16i1.1293

[36] Wiguna W. Decision tree of Coronavirus Disease (COVID-19) surveillance. IEEE Dataport. 2020. DOI: 10.21227/remc-6d63

Chapter 3

Association Rule Mining on Big Data Sets

Oguz Celik, Muruvvet Hasanbasoglu, Mehmet S. Aktas and Oya Kalipsiz

Abstract

An accurate, complete, and rapid establishment of customer needs and existence of product recommendations are crucial points in terms of increasing customer satisfaction level in various different sectors such as the banking sector. Due to the significant increase in the number of transactions and customers, analyzing costs regarding time and consumption of memory becomes higher. In order to increase the performance of the product recommendation, we discuss an approach, a sample data creation process, to association rule mining. Thus instead of processing whole population, processing on a sample that represents the population is used to decrease time of analysis and consumption of memory. In this regard, sample composing methods, sample size determination techniques, the tests which measure the similarity between sample and population, and association rules (ARs) derived from the sample were examined. The mutual buying behavior of the customers was found using a well-known association rule mining algorithm. Techniques were compared according to the criteria of complete rule derivation and time consumption.

Keywords: big data, sampling, association rule mining, data mining, data preprocessing techniques

1. Introduction

Thanks to improved storage capacities, databases in various fields such as banking have grown up to a rich level. Most of the strategic sales and marketing decisions are taken by processing these data. For example, strategies such as cross-sell, up-sell, or risk management are being created as a result of processing the customer data. Because of the increasing number of customers and the need for a higher processing capacity, it has made it more difficult to identify the customer requirements in a rapid and accurate way and to present solution recommendations. Innovative data mining applications and techniques are required to solve this issue [1].

The market basket analysis is one of the data mining methods applied to identify the pattern which is found in product ownership data of customers. Thanks to this analysis, a pattern among the products frequently bought together by the customers can be established. The obtained pattern plays an active role in developing cross-sell and up-sell strategies.

Market basket analysis consists of two main processes. These are clustering and association processes, respectively. The clustering process involves grouping of similar customers in terms of clusters. Thus, those customers which should be examined in the same category will be identified. During the association process, commonness in buying behavior of customers through a selected cluster is being identified, assuming that clustered customers having similar characteristics would demonstrate similar buying behaviors.

As the banking databases have grown up to a very high volume, the association process has become a very costly process in terms of time and memory consumption. In order to improve the time and memory performance, sampling process should be included in the previous phase of association.

In this regard, a sample which involves less observations in comparison to the whole data is used. We use the term "space" to refer the whole data set. In case the representation capability of the obtained sampling is high, loss of data is minimized, and the association process is realized through the sample instead of the space itself. Thus, less data shall be processed, and association rules (ARs) shall be obtained faster by consuming less memory.

As the subject of this book chapter was focused onto the banking data, customer segmentation conducted by the bank data was accepted as the clustering. As a result of the segmentation, clusters created by similar customers were used as input of sampling.

In this chapter, sample creation methods, techniques to find ideal sampling size, the space representing capability of these samples generated by these techniques, and association rules discovered through these samples were examined, respectively. Association rules obtained from both the space and sample were used to verify the sampling process. Besides, the spared amount in terms of time consumption was calculated.

This book chapter was organized as follows: Section 2 explains the studies toward deriving association rules through the space and sampling. Section 3 explains the parameters required to obtain association rules and the Apriori algorithm. Section 4 contains parameters to create the sample, sample creation methods, and the techniques used to calculate the sample size. Section 5 examines association rules obtained from the space and the sample and the results showing the representation capability of the sample for the respective space and the results showing rewards in terms of time consumption. Section 6 gives an overview and concludes the chapter.

2. Related work

In association rule mining, first the item sets, which are found together frequently, are found, and then the rules are obtained from these item sets.

Association algorithms are classified according to characteristics of the obtained item sets. In early studies Agrawal-Imielinski-Swami (AIS) algorithm which was allowed to find wide item sets was used, and then algorithms were found such as Apriori, which were used frequently now and which were able to process the bigger data sets faster [2].

The mutual usage of association discovery and sample creation methods is not a new approach. Sample creation studies toward association detection have begun with papers demonstrating mathematically that it was possible to create a sample which maintained the characteristics of the space. The following studies involved several techniques calculating the optimal number of observations [3–7].

At the beginning of the sample size detection studies, the data to be sampled were not considered; they have tried to determine the sample size using parameters not depending on the data such as margin of error, minimum support, and minimum confidence [3]. In current studies, formulas (using variables such as maximal process length or Vapnik-Chervonenkis (VC) size of the data cluster) considering the data characteristics have appeared [4–7]. There exists a number of studies focusing on how the management of metadata of big data sets are provided in a distributed computing setting [8–11]. Moreover, there exists a number of studies that are conducted in the field of information systems for managing distributed data storage platforms [12–16]. Unlike these studies, this chapter focuses on extracting meaningful information, i.e. association rules, from the big data sets. Initial results of the experimental studies, covered in this chapter, were reported in a previous study [17]. Sections 3 and 4 give detailed information on association detection and sample creation methods, respectively, and explain the techniques used in this study.

3. Association detection methods

In data mining, it is used to determine the pattern found among the association algorithms and observations [2, 18, 19]. In case any organization's transaction database is discussed, an analogy can be established between the observations and customers and between areas where a pattern is tried to be found and the bought products. Patterns obtained by association algorithms are processed to obtain association rules.

Association rules may be defined as follows: let us call each subset of products within the database an "itemset," and let us call each set of products purchased together by the customer a "transaction." The support count of any itemset is defined as the number of transactions associated with the items in the set within the database. The support indicates the ratio of support count to the number of transactions within the database. The itemset which meets the minimum support requirement is called the frequent itemset (FI).

For example, if a database with 10 transactions contains product A in 3 different transactions, then the product A's support count is 3, and its support is 0.3. In case the minimum support is defined by a value lower than 0.3, then the product A will be classified as FI.

There are several algorithms deriving FI using the transactions within the database [2, 18]. In this chapter, Apriori algorithm was preferred due its ability of deriving all itemsets within the space. This algorithm derives primarily candidate itemsets starting with one-element itemset from the database. Those providing minimum support from candidate itemsets are filtered and recorded as FI. New candidate itemsets are created from the FI obtained in the previous step by increasing the number of elements. In each step, the candidate itemsets are passed through a minimum support test, and the algorithm continues until no FI with k-elements can be generated.

Among the elements of the FI obtained from the database, it is possible to derive association rules in A- > B format. Then, AR's support gets equal to AUB itemset support. The confidence is defined as the ratio of AUB itemset support to the A itemset support. AR should meet the minimum confidence requirement specified by the customer [2].

Assuming that A - > B rule has a support of s and the confidence of c, we can derive that the itemsets A and B in the whole database are associated with a probability of s and a customer owning the itemset A might be an owner of the itemset B with a probability of c.

To find out all ARs within the database, a rule mining algorithm is applied to each FI obtained. Candidate rule combinations are created for rule mining among all subsets of a selected FI in A - > B format. Those providing minimal confidence from candidate rules are filtered and recorded as association rule.

4. Sample creation methodology

Sample creation is the process of creating a subset containing the characteristics of a data set. The subset created through sampling is expected to represent the data set (space). In traditional statistical methods, the similarity of two data sets is measured by either $\chi 2$ test or Kolmogorov–Smirnov (K-S) test.

In this study, these tests were utilized, in order to measure the similarity of the created sample in comparison with its space. A comparison was conducted through p values (the probability p of finding the space characteristics) of the statistics resulting from both tests. In case the obtained p value exceeds 0.05, it can be deducted that "the sample is similar to the space with a probability of at least 95%."

Sample creation is discussed under two topics, i.e., sample creation methods and sample size determination techniques. Sample creation methods are explained in Chapter 4.1 and sample size determination techniques explained in Chapter 4.2.

4.1 Sample creation methods

When creating samples from the space, it is possible to use several sample creation methods. These methods are classified according to the selection of observations from the space. The main sample creation methods are as follows:

4.1.1 Simple random sampling

The observations within the space are selected without following a specific routine. The selection probability of each observation is equal.

Systematic sampling: The observations within the space are numbered. Sampling interval is created by dividing the space size to the observation size. A random number is selected. The observation sample at this number from each interval is included.

4.1.2 Stratified sampling

This is used where the observations within the space can be divided into groups. The samples are created maintaining the ratio between the number of observations of groups within the space and the total number of observations. The selection probability of each observation in the same stratus is equal.

4.1.3 Cluster sampling

This is used where the observations within the space can be divided into groups. After the groups are determined, they are selected using the simple random sampling method. All observations within selected groups are included into the sample.

4.1.4 Multistage sampling

This is used where the observations within the space can be divided into groups. After the groups are determined, groups are selected by the simple random sampling method. Unlike cluster sampling, observations to be selected from groups are determined by the simple random sampling method.

Among the mentioned methods, the simple random sampling method stands up by its high speed. As the methods, which require creation of groups within the space and sorting of observations, need a pre-analysis, their time consumption is more than the simple random sampling method.

4.2 Sample size determination methods

The expected parameter in sample creation methods is the size of the sample to be created. When the optimal sample size is calculated, a number which will not decrease its space representing capability should be found. Under association detection algorithms, it is important to derive all FIs and ARs within the space from the sample. In this study, techniques specialized on association detection algorithms have been examined from those developed for sample size determination [3–5, 7]. Sample size determination techniques are divided into two groups to minimize the FI and AR loss.

When the association algorithms will be run using the same parameters, support and confidence values calculated from the sample appear to be different than their counterparts calculated from the space. This margin of error is measured using two different methods. When calculating absolute margin of error, the absolute value of the difference between values from the space and the sample is considered. The relative margin of error is calculated by dividing absolute margin of error into the value within the space. In **Table 1**, the lines containing "absolute" at the "type of technique" column aim at reducing absolute margin of error, while those containing "relative" aim to minimize relative margin of error.

All examined techniques are shown in **Table 1** with suggested formulas and type of formula. The values found through the techniques determine the minimum number of transactions required for sample creation. The number of transactions which are equal to the values found is selected from the space by the preferred sample creation method.

Sample size determination techniques determine the minimum number of transactions required for sample creation. The number of transactions which are equal to the values found is selected from the space by the preferred sample creation method.

The complexity of the space is calculated theoretically using the Vapnik-Chervonenkis size [20]. Assuming that the transactions within the database are sorted according to their number of elements and that the "number of transactions" and "number of elements" are plotted on the coordinate system, the d-index value would correspond to the edge length of the largest square.

Description	Type of Technique	Formula
Zaki	FI-absolute	$\frac{-2\ln(1-\gamma)}{\Theta\delta^2}$
Toivonen	FI-absolute	$\frac{1}{2\varepsilon^2}\ln\frac{2}{\delta}$
Chakaravarthy	FI-absolute	$\frac{24}{(1-\varepsilon)\varepsilon^2\Theta}\left(\Delta+5+\ln\frac{4}{(1-\varepsilon)\Theta\delta}\right)$
Chakaravarthy	AR-absolute	$\frac{48}{(1-\varepsilon)\varepsilon^2\Theta}\left(\Delta+5+\ln\frac{5}{(1-\varepsilon)\Theta\delta}\right)$
Riondato	FI-absolute	$\frac{4c}{\varepsilon^2}\left(v+\ln\frac{1}{\delta}\right)$
Riondato	FI-relative	$\frac{4(2+\varepsilon)c}{\varepsilon^2(2-\varepsilon)\Theta}\left(v\ln\frac{2+\varepsilon}{\Theta(2-\varepsilon)}+\ln\frac{1}{\delta}\right)$
Riondato	AR-absolute	$\frac{c}{\eta^2 p}\left(v\ln\frac{1}{p}+\ln\frac{1}{\delta}\right)$
Riondato	AR-relative	$\frac{c}{\eta^2 p}\left(v\ln\frac{1}{p}+\ln\frac{1}{\delta}\right)$

Minimal sample size can be determined in terms of accuracy ε, probability of error δ, minimum support Θ, minimum confidence γ, d-index value of the space v, maximal process length of the space Δ, and the constant c. In formulas, the value η is calculated depending on variables Θ, γ, and ε; and the value p is calculated depending on values η and Θ.

Table 1.
Sample size calculation techniques are provided.

In this study, we use a d-index algorithm, which does not seek a sorting requirement among the transactions, and it calculates v (d-index value), by initializing with 1 and by increasing it. All transactions within the database should be scanned to find the value. Where a number of transactions are large and the item number within each transaction is less, such as banking data, the length of transactions becomes decisive in determining the d-index value. In d-index algorithm, the transactions within the database were sorted in descending order of item numbers, and the value v was calculated decreasingly beginning from the maximal transaction length. Here, it is not necessary to scan all transactions.

5. Experimental evaluation and results

The tests were performed on product ownership data of banking customers. Statistical studies' code development was performed on the widely used R programming language.

When tests were performed, the steps below were followed:

1. Determine the sample size utilizing various techniques

2. Create three different samples for each technique using the simple random sampling method

3. Compare the representability of the space for the obtained sample examination with $\chi 2$ and K-S tests

4. Use the Apriori algorithm included in the *arules* package of R language, and determine the FI and AR through the space and sample

5. Calculate the absolute error in support and trust values, and compare the results with those obtained from the space

6. Compare the duration of obtaining AR and the duration of sample creation. Generate AR from the sample

Theoretically, it is expected that the samples in various sizes obtained from FI and AR results are tuned with the results from the space, that there is a correspondence between representability and absolute error, and that the duration of transactions made on the sample and the memory consumption reduce.

To accelerate the test processes, instead of 143 products of the bank, 10 different product groups were determined, and the association between those groups was examined. The utilized banking data is a matrix including 1,048,575 customers and an ownership status of customers about 10 different product groups. The lines represent customers and the columns represent product groups. In case the customer owns a product, the intersection of that line-column indicates 1, otherwise 0. In these tests the following parameters were used: accuracy $\varepsilon = 0.04$, probability of error $\delta = 0.07$, minimal support value $\Theta = 0.02$, and minimum confidence $\gamma = 0.06$, 0.1, and 0.14.

Table 2 shows varying sample sizes corresponding to varying minimum confidence values. Because γ was not used as a parameter in formulas Toivonen, Chakaravarthy FI-absolute, Chakaravarthy AR-absolute, Riondato FI-absolute, and Riondato FI-relative, there are no variations in calculated sizes.

When **Table 2** was examined in detail, it is obvious that the sizes obtained from the techniques Chakaravarthy FI-absolute, Chakaravarthy AR-absolute, Riondato

Description	Type of Technique	$\gamma = 0.06$	$\gamma = 0.1$	$\gamma = 0.14$
Zaki	FI-absolute	3867	6585	9426
Toivonen	FI-absolute	1047	1047	1047
Chakaravarthy	FI-absolute	14842499	14842499	14842499
Chakaravarthy	AR-absolute	30033660	30033660	30033660
Riondato	FI-absolute	9574	9574	9574
Riondato	FI-relative	1458404	1458404	1458404
Riondato	AR-absolute	15057	47005	96859
Riondato	AR-relative	5468750	5468750	5468750

Techniques where the calculated size is larger than the space were not used at the sample creation step.

Table 2.
Calculated sample sizes based on varying minimum trust values are provided.

FI-relative, and Riondato AR-relative are larger than the space (1,048,575). As the aim was to reduce the data set, these techniques were not examined in the following tests. In order to minimize the error due to simple random sampling method, three samples were created for each of the Zaki, Toivonen, Riondato FI-absolute, and Riondato AR-absolute techniques.

Table 3 shows average *p* values calculated from $\chi 2$ and K-S tests. As the similarity significance between the space and sample was accepted as 95%, it is expected that p values are higher than 0.05. The results indicate that values were obtained to prove an adequate statistical similarity between the space and all obtained samples. An instability is obvious regarding *p* values of Toivonen where the sample size does not vary. We consider this instability results from the small sample size provided by this technique.

FIs and their corresponding ARs were determined from the samples created using Apriori algorithm. To measure the similarities of FIs and ARs, absolute error was calculated through support and trust values. Zaki and Toivonen techniques were inadequate to determine all FIs and ARs existing in the space for the value γ = 0.1. Because a loss of rule was undesirable, we have observed that these two techniques were not suitable to sample creation and time consumption tests were not examined. By calculating the error by substituting incomplete values with 0, the results on **Table 4** were obtained. As expected, where absolute support error was high, an also high-confidence error was found.

In a comparative review of **Tables 3** and **4**, no relation was detected between support and confidence errors by the results obtained from $\chi 2$ and K-S tests. We have noticed that traditional statistical measurements were inadequate in measuring the representability of the sample which was created for association mining.

In **Table 5**, durations until creation of AR are provided for the space and created samples. While the duration from sample creation until obtaining the AR was provided for the space, the time required for sample size determination, total average time required for sample creation, and obtaining the AR by simple random

Description	Type of Technique	$\gamma = 0.06$		$\gamma = 0.1$		$\gamma = 0.14$	
		χ^2	K-S	χ^2	K-S	χ^2	K-S
Zaki	FI-absolute	0.824	0.591	0.630	0.439	0.190	0.382
Toivonen	FI-absolute	0.379	0.512	0.435	0.395	0.341	0.675
Riondato	FI-absolute	0.142	0.182	0.595	0.434	0.234	0.081
Riondato	AR-absolute	0.690	0.618	0.663	0.300	0.385	0.396

All the techniques were found to be similar to the space.

Table 3.
p values calculated from $\chi 2$ and K-S tests were provided based on minimum trust values.

Description	Type of Technique	γ = 0.06	γ = 0.1	γ = 0.14
		Supp. Conf.	Supp. Conf.	Supp. Conf.
Zaki	FI-absolute	0.002 0.009	0.003 0.06	0.002 0.001
Toivonen	FI-absolute	0.005 0.022	0.006 0.042	0.004 0.001
Riondato	FI-absolute	0.023 0.008	0.002 0.011	0.002 0.001
Riondato	AR-absolute	0.011 0.004	0.001 0.001	0.001 0.001

Techniques where AR loss was experienced were not tested in terms of running time.

Table 4.
Average support and trust absolute error generated based on varying minimum trust values are provided.

Description	Type of Technique	γ = 0.06	γ = 0.1	γ = 0.14
Space	–	1,832	1,825	1.87
Riondato	FI-absolute	0,186	0,193	0,193
Riondato	AR-absolute	0,193	0,253	0,343

Techniques where AR loss was experienced were not tested in terms of running time.

Table 5.
The time until rule mining based on varying minimum trust values γ was given in seconds.

sampling method were provided for the samples. As expected, the time performance of all techniques for each value γ was found better than the space. Even more benefits are expected by using actually 143 different products instead of 10 product groups in these tests.

FI and AR results of the samples were compared to those generated from the space. Within the space, the mostly encountered depository (D) product has a ratio of 94%, credit card (CC) product has 11%, and installment loan (IL) product has 8%. These products are found together within the space by a ratio of 2.3%. In spite of this low support value, three different rules with high confidence were derived. "CC, IL → D" rule was derived, and eventually, it was observed that 92% of people who have bought CC and IL also have bought D. According to another derived rule "IL, D -> CC," it was discovered that 31% of people who bought CC and D also bought IL. It was also found through another rule CC, D - > IL that 28% of people who bought IL and D also bought CC. These rules could not be derived from the techniques Zaki and Toivonen, and information loss was experienced. Therefore, these techniques are not suitable on sample creation for association mining. As expected, when sample size increased, the obtained absolute error in results decreased.

The utilized banking data contains information about customers and the product groups of their owned products. In other words, instead of products owned by customers, banking product groups were used. This was preferred to reduce the data set sparsity and to accelerate the test speeds. So, it was observed that even tests on the space did not take longer than 2 seconds. Whenever the advantages in terms of duration seem to be in the order of seconds, tests to be conducted with a data set containing more products (or product groups) will show decisive advantages.

6. Conclusion and future work

Because AR mining process through the space takes a long time, we have aimed at determining a smaller sized sample representing the space and AR mining through that sample. For this purpose, in this book chapter, we have investigated for techniques which provide an ideal sample size specialized on association mining.

The samples were created using the simple random sampling method, and three different samples were obtained per technique. We have tried to prevent a potential noise in our results by creating multiple samples.

The similarity of samples to the space was measured by the $\chi 2$ test and K-S test. It was obvious that after both tests the obtained values for association mining were inadequate in measuring the representability of samples. In those tests, no relationship was found among support and confidence error values. We consider that the probability of tests giving biased outputs and the inadequacy of suggested sample sizes in measuring were the reasons for having these results.

The results indicate that the duration of AR generation within the space was compared to the total time of sample size determination, sample creation, and AR generation through the sample. It was observed that each technique was better performing in terms of space results. Riondato FI-absolute and Riondato AR-absolute techniques have given good results based on calculated absolute error values. When smaller sample size and less time consumption criteria were considered, Riondato FI-absolute technique becomes favorable.

In future studies, the data set shall be renewed in this regard, and other sample methods will also be applied. Besides, results which might be related to a single data set shall be extended with tests to be performed on another data set, and the results shall be cross-checked.

Author details

Oguz Celik, Muruvvet Hasanbasoglu, Mehmet S. Aktas* and Oya Kalipsiz
Department of Computer Engineering, Yildiz Technical University Istanbul, Turkey

*Address all correspondence to: aktas@yildiz.edu.tr

References

[1] Jayasree V, Balan RVS. A review on data mining in banking sector. American Journal of Applied Sciences. 2013;**10**(10):1160-1165

[2] Agrawal R, Imieliński T, Swami A. Mining association rules between sets of items in large databases. In: SIGMOD '93, Proceedings of the 1993 ACM SIGMOD International Conference on Management of Data; 1993

[3] Zaki MJ, Parthasarathy S, Li W, Ogihara M. Evaluation of sampling for data mining of association rules. In: Proceedings 7th International Workshop on Research Issues in Data Engineering; 1997

[4] Chakaravarthy TV et al. Analysis of sampling techniques for association rule mining. In: 12th International Conference on Database Theory; 2009

[5] Klemettinen M et al. Finding interesting rules from large sets of discovered association rules. In: CIKM '94: Proceedings of the 3rd International Conference on Information and Knowledge Management; 1994

[6] Toivonen H. Sampling large databases for association rules. In: Proceedings of the 22nd VLDB Conference, Mumbai (Bombay), India. 1996. pp. 134-145

[7] Riondato M et al. Efficient discovery of association rules and frequent item sets through sampling with tight performance guarantees. ECML PKDD. 2012:25-41

[8] Baeth MJ et al. An approach to custom privacy violation detection problems using big social provenance data. Concurrency and Computation-Practice & Experience. 2018;**30**(21):1-9

[9] Baeth MJ et al. Detecting misinformation in social networks using provenance data. Concurrency and Computation-Practice & Experience. 2019;**31**(3):1-13

[10] Riveni M et al. Application of provenance in social computing: A case study. Concurrency and Computation-Practice & Experience. 2019;**31**(3):1-13

[11] Tas Y et al. An approach to standalone provenance systems for big provenance data. In: The International Conference on Semantics, Knowledge and Grids on Big Data (SKG-16); 2016. pp. 9-16

[12] Aktas MS. Hybrid cloud computing monitoring software architecture. Concurrency and Computation: Practice and Experience. 2018;**30**(21):1-9

[13] Aktas MS et al. A web based conversational case-based recommender system for ontology aided metadata discovery. In: The 5th IEEE/ACM International Workshop on Grid Computing. 2004. pp. 69-75

[14] Aktas MS et al. Fault tolerant high-performance information services for dynamic collections of Grid and Web services. Future Generation Computer Systems. 2007;**23**(3):317-337

[15] Pierce ME et al. The QuakeSim project: Web services for managing geophysical data and applications. Pure and Applied Geophysics. 2008;**165**(3-4):635-651

[16] Aydin G et al. SERVOGrid complexity computational environments (CCE) integrated performance analysis. In: The 6th IEEE/ACM International Workshop on Grid Computing; 2005. pp. 256-261

[17] Celik O et al. Implementation of data preprocessing techniques on

distributed big data platforms. In: 4th International Conference on Computer Science and Engineering (UBMK); 2019. pp. 73-78

[18] Eltabakh MY et al. Incremental mining for frequent patterns in evolving time series databases. Technical report. Department of Computer Science, Purdue University; 2008

[19] Pei J, Han J, Lu H, Nishio S, Tang S, Yang D. H-Mine: Fast and space-preserving frequent pattern mining in large databases. IIE Transactions. 2007;**39**(6):593-605. DOI: 10.1080/07408170600897460

[20] Vapnik V et al. Measuring the VC-dimension of a learning machine. Neural Computation. 1994;**6**(5):851-876

Chapter 4

Data Mining in Banking Sector Using Weighted Decision Jungle Method

Derya Birant

Abstract

Classification, as one of the most popular data mining techniques, has been used in the banking sector for different purposes, for example, for bank customer churn prediction, credit approval, fraud detection, bank failure estimation, and bank telemarketing prediction. However, traditional classification algorithms do not take into account the class distribution, which results into undesirable performance on imbalanced banking data. To solve this problem, this paper proposes an approach which improves the decision jungle (DJ) method with a class-based weighting mechanism. The experiments conducted on 17 real-world bank datasets show that the proposed approach outperforms the decision jungle method when handling imbalanced banking data.

Keywords: data mining, classification, banking sector, decision jungle, imbalanced data

1. Introduction

Data mining is the process of analyzing large data stored in data warehouses in order to automatically extract hidden, previously unknown, valid, interesting, and actionable knowledge such as patterns, anomalies, associations, and changes. It has been commonly used in a wide range of different areas that include marketing, health care, military, environment, and education. Data mining is becoming increasingly important and essential for banking sector as well, since the amount of data collected by banks has grown remarkably and the need to discover hidden and useful patterns from banking data becomes widely recognized.

Banking systems collect huge amounts of data more rapidly as the number of channels (i.e., Internet banking, telebanking, retail banking, mobile banking, ATM) has increased. Banking data has been currently generated from various sources, including but not limited to bank account transactions, credit card details, loan applications, and telex messages. Hence, data mining can be used to extract meaningful information from these collected banking data, to enable banking institutions to make better decision-making process. For example, *classification*, which is one of the most popular data mining techniques, can be used to predict bank failures [1–3], to estimate bank customer churns [4], to detect frauds [5], and to evaluate loan approvals [6].

In many real-world banking applications, the distribution of the classes in the dataset is highly skewed. A bank data is *imbalanced*, when its target variable is categorical and if the number of samples in one class is significantly different from those of the other class(es). For example, in credit card fraud detection, most of the instances in the dataset are labeled as "non-fraud" (majority class), while very few are labeled as "fraud" (minority class). Similarly, in bank customer churn prediction, many instances are represented as negative class, whereas the minorities are marked as positive class. However, the performance of classification models is significantly affected by a skewed distribution of the classes; hence, this imbalance problem in the dataset may lead to bad estimates and misclassifications. Dealing with imbalanced data has been considered as one of the 10 most difficult problems in the field of data mining [7]. With this motivation, this paper proposes a class-based weighting strategy.

The main contribution of this paper is that it improves the decision jungle (DJ) method by a class-based weighting mechanism to make it effective in handling imbalanced data. In the proposed approach, a weight is assigned to each class based on its distribution, and this weight value is combined with class probabilities. The experimental studies conducted on 17 real-world banking datasets confirm that our approach generally performs better than the traditional decision jungle algorithm when the data is imbalanced.

The rest of this paper is organized as follows. Section 2 briefly presents the recent and related research in the literature. Section 3 describes the proposed approach, class-based weighted decision jungle method, in detail. Section 4 is devoted to the presentation and discussion of the experimental results, including the dataset descriptions. Finally, Section 5 gives the concluding remarks and provides some future research directions.

2. Related work

As a data-intensive sector, banking has been a popular application area for data mining researchers since the information technology revolution. The continuous developments in banking systems and the rapidly increasing availability of big banking data make data mining one of the most essential tasks for the banking industry.

Banking industries have used data mining techniques in various applications, especially on bank failure prediction [1–3], possible bank customer churns identification [4], fraudulent transaction detection [5], customer segmentation [8–10], predictions on bank telemarketing [11–14], and sentiment analysis for bank customers [15]. Some of the classification studies in the banking sector have been compared in **Table 1**. The objectives of the studies, years they were conducted, algorithms and ensemble learning techniques they used, the country of the bank, and obtained results are shown in this table.

The main data mining tasks are classification (or categorical prediction), regression (or numeric prediction), clustering, association rule mining, and anomaly detection. Among these data mining tasks, classification is the most frequently used one in the banking sector [16], which is followed by clustering. Some banking applications [8, 10] have used more than one data mining techniques, among which clustering before classification has shown sufficient evidence of both popularity and applicability.

Apart from novel task-specific algorithms proposed by the authors, the most commonly used classification algorithms in the banking sector are decision tree (DT), neural network (NN), support vector machine (SVM), k-nearest neighbor

Ref	Year	Algorithms						Ensemble learning		Description	Country of the bank	Result
		DT	NN	SVM	KNN	NB	LR	Bagging (i.e., RF)	Boosting (AB, XGB)			
Manthoulis et al. [1]	2020			√			√		√	Bank failure prediction	USA	AUC >0.97
Ilham et al. [11]	2019	√	√	√	√	√	√	√		Long-term deposit prediction	Portugal	ACC 97.07%
Lv et al. [5]	2019		√							Fraud detection in bank accounts	—	ACC 97.39%
Krishna et al. [15]	2019	√	√	√	√	√	√	√	√	Sentiment analysis for bank customers	India	AUC 0.8268
Farooqi and Iqbal [12]	2019	√	√	√	√	√				Prediction of bank telemarketing outcomes	Portugal	ACC 91.2%
Carmona et al. [2]	2019						√	√	√	Bank failure prediction	USA	ACC 94.74%
Jing and Fang [3]	2018		√	√			√			Bank failure prediction	USA	AUC 0.916
Lahmiri [13]	2017		√							Prediction of bank telemarketing outcomes	Portugal	ACC 71%
Marinakos and Daskalaki [8]	2017	√	√	√	√		√			Customer classification for bank direct marketing	Portugal	AUC 0.9
Keramati et al. [4]	2016	√								Bank customer churn prediction	—	AUC 0.929
Wan et al. [6]	2016	√		√	√			√	√	Predicting nonperforming loans	China	AUC 0.965
Ogwueleka et al. [10]	2015		√				√			Identifying bank customer behavior	Intercontinental	AUC 0.94
Moro et al. [14]	2014	√	√	√			√			Prediction of bank telemarketing outcomes	Portugal	AUC 0.8
Smeureanu et al. [9]	2013		√	√						Customer segmentation in banking sector	Romania	ACC 97.127%

Table 1.
Classification applications in the banking sector.

(KNN), Naive Bayes (NB), and logistic regression (LR), as shown in **Table 1**. Some data mining studies in the banking sector [1, 2, 6, 11, 15] have used ensemble learning methods to increase the classification performance. Bagging and boosting are the most popular ensemble learning methods due to their theoretical performance advantages. Random forest (RF) [2, 6, 11, 15], AdaBoost (AB) [6], and extreme gradient boosting (XGB) [2, 15] have also been used in the banking sector as the most well-known bagging and boosting algorithms, respectively. As shown in **Table 1**, accuracy (ACC) and area under ROC curve (AUC) are the commonly used performance measures for classification.

Dealing with class imbalance problem, various solutions have been proposed in the literature. Such methods can be mainly grouped under two different approaches: (i) application of a data preprocessing step and (ii) modifying existing methods. The first approach focuses on balancing the dataset, which may be done either by increasing the number of minority class examples (over-sampling) or reducing the number of majority class examples (under-sampling). In the literature, synthetic minority over-sampling technique (SMOTE) [17] is commonly used as an over-sampling technique. As an alternative approach, some studies (i.e., [18]) focus on modifying the existing classification algorithms to make them more effective when dealing with imbalanced data. Unlike these studies, this paper proposes a novel approach (class-based weighting approach) to solve imbalanced data problem.

3. Methods

3.1 Decision jungle

A *decision jungle* is an ensemble of rooted decision *directed acyclic graphs* (DAGs), which are powerful and compact distinct models for classification. While a traditional decision tree only allows one path to every node, a DAG in a DJ allows multiple paths from the root to each leaf [19]. During the training phase, node splitting and merging operations are done by the minimization of an objective function (the weighted sum of entropies at the leaves).

Unlike a decision forest that consists of several evolutionary induced decision trees, decision jungle consists of an ensemble of decision directed acyclic graphs. Experiments presented in [19] show that decision jungles require significantly less memory while significantly improving generalization, compared to decision forests and their variants.

3.2 Class-based weighted decision jungle method

In this study, we improve the decision jungle method by a class-based weighting mechanism to make it effective in dealing with imbalanced data.

Giving a training dataset $D = \{(x_1, y_1), (x_2, y_2), \ldots, (x_n, y_N)\}$ that contains N instances, each instance is represented by a pair (x, y), where x is a d-dimensional vector such that $x_i = [x_{i1}, x_{i2}, \ldots, x_{id}]$ and y is its corresponding class label. While x is defined as input variable, y is referred as output variable in the categorical domain $Y = \{y_1, y_2, \ldots, y_k\}$, where k is the number of class labels. The goal is to learn a classifier function $f: X \rightarrow Y$ that optimizes some specific evaluation metric(s) and can predict the class label for unseen instances.

Training dataset is usually considered as a set of samples from a probability distribution F on $X \times Y$. An instance component x is associated with a label class y_j of Y such that:

$$\frac{P\left(y_j|x\right)}{P\left(y_m|x\right)} > threshold, \forall m \neq j \tag{1}$$

where $P(y_j\,|x)$ is the predicted conditional probability of x belonging to y_j and threshold is typically set to 1.

In this paper, we focus on imbalanced data problem, where the number of instances in one class (y_i) is much larger or less than instances in the other class (y_j). Like many other classification algorithms, the decision jungle method is also affected by a skewed distribution of the classes, because the traditional classifiers tend to be overwhelmed by the majority class and ignore the rare samples in the minority class. In order to overcome this problem, we locally adapted a class-based weighted mechanism, where weights are determined depending on the distribution of the class labels in the dataset. The main idea is that the minority class receives a higher weight, while the majority class is assigned with a lower weight during the combination class probabilities. According to this approach, the weight over a class is calculated as follows:

$$W_c = \frac{\frac{1}{Log(N_c+1)}}{\sum_{i=1}^{k} \frac{1}{Log(N_i+1)}} \tag{2}$$

where W_c is the weight assigned to the class c, N is the total number of instances in the dataset, N_c is the number of instances present in the class c, and k is the number of class labels. In the proposed approach, Eq. (1) is updated as follows:

$$\frac{W_j * P\left(y_j|x\right)}{W_m * P\left(y_m|x\right)} > threshold, \forall m \neq j \tag{3}$$

Figure 1 shows the general structure of the proposed approach. In the first step, various types of raw banking data are obtained from different sources such as account transactions, credit card details, loan applications, and social media texts. Next, raw banking data is preprocessed by applying several different techniques to provide data integration, data selection, and data transformation. The prepared data is then passed to the training step, where weighted decision jungle algorithm is used to build an effective model which accurately maps inputs to desired outputs. The classification validation step provides feedback to the learning phase for adjustment

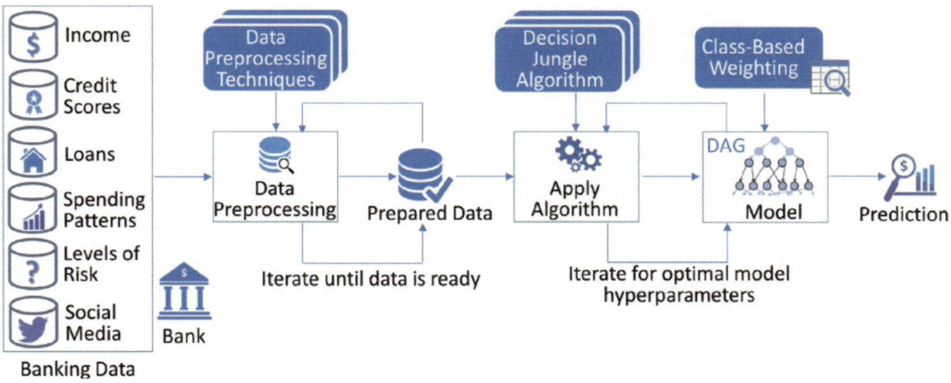

Figure 1.
General structure of proposed approach.

to improve model performance. The training phase is repeated until a desired classification performance is achieved. Once a model is build, after that it can be used to predict unseen data.

4. Experimental studies

We implemented the proposed approach in Azure Machine Learning Studio framework on cloud platform. In all experiments, default input parameters of the decision forest algorithm were used as follows:

- Ensemble approach: Bagging
- Number of decision DAGs: 8
- Maximum width of the decision DAGs: 128
- Maximum depth of the decision DAGs: 32
- Number of optimization steps per decision DAG layer: 2048

Conventionally, *accuracy* is the most commonly used measure for evaluating a classifier performance. However, in the case of imbalanced data, accuracy is not sufficient alone since the minority class has very little impact on accuracy than the majority class. Using only accuracy measure is meaningless when the data is imbalanced and where the main learning target is the identification of the rare samples. In addition, accuracy does not distinguish between the numbers of correct class labels or misclassifications of different classes. Therefore, in this study, we also used several more metrics: *macro-averaged precision*, *recall*, and *F-measure*.

4.1 Dataset description

In this study, we conducted a series of experiments on 17 publically available real-world banking datasets which are described in **Table 2**. We obtained eight from the UCI Machine Learning Repository [20] and nine datasets from Kaggle data repository.

4.2 Experimental results

Table 3 shows the comparison of the classification performances of DJ and weighted DJ methods. According to the experimental results, on average, the weighted DJ method shows better classification outcome than its traditional version on the imbalanced banking datasets in terms of both accuracy and recall metrics. For example, the imbalanced dataset "bank additional" has an accuracy of 94.54% with the DJ method and 94.61% with the weighted DJ method. The accuracy is slightly higher with the weighted version because the classifier was able to classify the minority class samples better (0.8385, instead of 0.7914). The proposed method only disappointed in its accuracy and recall values for 4 of 17 datasets (with IDs 5, 9, 12, and 13).

It is observed from the experiments that the weighted DJ method failed in classifying only one dataset among 17 datasets in terms of macro-averaged recall values. This means that the proposed method generally can be able to build a good model to predict minority class samples.

No	Dataset		#Instances	#Features	#Class	Majority class (%)	Minority class (%)	Data source
1	Abstract dataset for credit card fraud detection		3075	12	2	85.4	14.6	Kaggle
2	Bank marketing [14]	Bank	4521	17	2	88.5	11.5	UCI
3		Bank full	45,211	17	2	88.3	11.7	UCI
4		Bank additional	4119	21	2	89.1	10.9	UCI
5		Bank additional full	41,188	21	2	88.7	11.3	UCI
6	Bank customer churn prediction		10,000	14	2	79.6	20.4	Kaggle
7	Bank loan status		100,000	19	2	77.4	22.6	Kaggle
8	Banknote authentication		1372	5	2	55.5	44.5	UCI
9	Credit approval		690	16	2	55.5	44.5	UCI
10	Credit card fraud detection [21]		284,807	31	2	99.8	0.2	Kaggle
11	Default of credit card clients [22]		30,000	25	2	77.9	22.1	UCI
12	German credit		1000	21	2	70.0	30.0	UCI
13	Give me some credit		150,000	12	2	93.3	6.7	Kaggle
14	Loan campaign response		20,000	40	2	87.4	12.6	Kaggle
15	Loan data for dummy bank		887,379	30	2	92.4	7.6	Kaggle
16	Loan prediction		614	13	2	68.7	31.3	Kaggle
17	Loan repayment prediction		9578	14	2	84.0	16.0	Kaggle

Table 2.
The main characteristics of the banking datasets.

It can be deduced from the average precision and recall values that higher classification rates can be achieved with the weighted DJ method for minority classes, while more misclassified points in majority classes may also be detectable in the case of imbalanced data.

Figure 2 shows the comparison of the classification performances of two methods in terms of F-measure: decision jungle and class-based weighted decision jungle (weighted DJ). In principle, F-measure is defined as $F = (2 \times \text{Recall} \times \text{Precision})/(\text{Recall} + \text{Precision})$, which is a harmonic mean between recall and precision. According to the results, for all banking datasets, the proposed method showed some increase or the same performance in the F-measure value.

It can be possible to conclude from the experiments that the minority and majority ratios are not the only issues in constructing a good prediction model. For example, the minority and majority ratios of the first and last datasets are very close, but the classification outcomes related to these datasets are not similar. Although the minority and majority class ratios are almost the same for these two datasets, there is a significant difference between the classification accuracy, precision, and recall values of the datasets, as can be seen in **Table 3**. There is also a need

ID	Dataset	Decision jungle			Class-based weighted decision jungle		
		Acc (%)	Precision	Recall	Acc (%)	Precision	Recall
1	Abstract dataset for credit card fraud detection	99.09	0.9918	0.9715	99.19	0.9923	0.9749
2	Bank	92.70	0.8909	0.7175	92.70	0.8492	0.7593
3	Bank full	91.06	0.8181	0.6874	91.17	0.8039	0.7217
4	Bank additional	94.54	0.9082	0.7914	94.61	0.8739	0.8385
5	Bank additional full	92.21	0.8332	0.7347	92.19	0.8126	0.7762
6	Bank customer churn prediction	87.37	0.8514	0.7291	87.40	0.8394	0.7411
7	Bank loan status	84.37	0.9170	0.6328	84.38	0.9169	0.6332
8	Banknote authentication	99.85	0.9987	0.9984	100.00	1.0000	1.0000
9	Credit approval	92.80	0.9273	0.9275	92.65	0.9257	0.9261
10	Credit card fraud detection	99.97	0.9915	0.9167	99.97	0.9861	0.9309
11	Default of credit card clients	83.05	0.7833	0.6695	83.16	0.7793	0.6785
12	German credit	86.30	0.8545	0.8088	85.70	0.8338	0.8198
13	Give me some credit	93.88	0.8245	0.5986	93.77	0.7861	0.6240
14	Loan campaign response	89.34	0.9393	0.5763	90.34	0.9390	0.6178
15	Loan data for dummy bank	95.19	0.9753	0.6837	95.20	0.9753	0.6844
16	Loan prediction	83.54	0.8715	0.7443	83.54	0.8631	0.7481
17	Loan repayment prediction	84.82	0.9059	0.5266	85.35	0.8900	0.5453
	Average	91.18	0.8990	0.7479	91.25	0.8863	0.7659

Table 3.
Comparison of unweighted and class-based weighted decision jungle methods in terms of accuracy, macro-averaged precision, and macro-averaged recall.

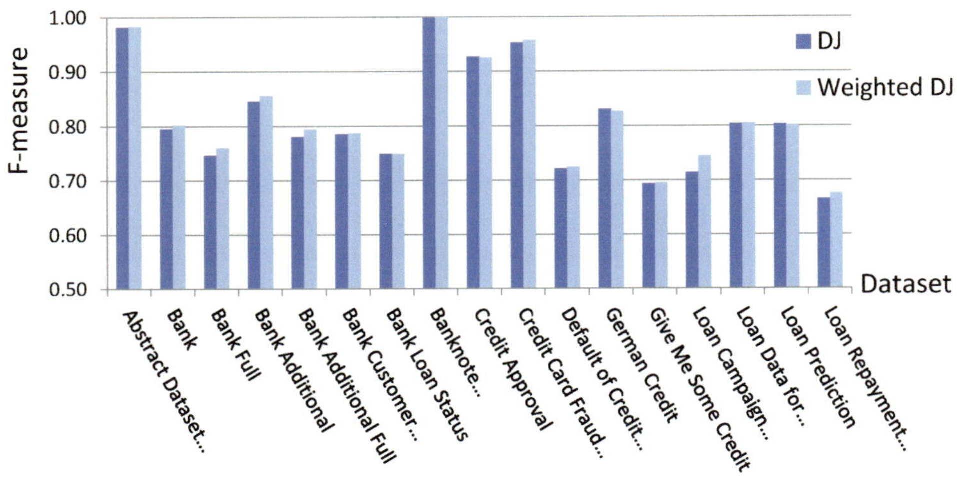

Figure 2.
Comparison of unweighted and class-based weighted decision jungle methods in terms of F-measure.

for appropriate training examples that have data characteristics consistent with the class label assigned to them.

5. Conclusion and future work

As a well-known data mining task, classification in real-world banking applications usually involves imbalanced datasets. In such cases, the performance of classification models is significantly affected by a skewed distribution of the classes. The data imbalance problem in the banking dataset may lead to bad estimates and misclassifications. To solve this problem, this paper proposes an approach which improves the decision jungle method with a class-based weighting mechanism. In the proposed approach, a weight is assigned to each class based on its distribution, and this weight value is combined with class probabilities. The empirical experiments conducted on 17 real-world bank datasets demonstrated that it is possible to improve the overall accuracy and recall values with the proposed approach.

As a future study, the proposed approach can be adapted for multi-label classification task. In addition, it can be enhanced for the ordinal classification problem.

Author details

Derya Birant
Department of Computer Engineering, Dokuz Eylul University, Izmir, Turkey

*Address all correspondence to: derya@cs.deu.edu.tr

References

[1] Manthoulis G, Doumpos M, Zopounidis C, Galariotis E. An ordinal classification framework for bank failure prediction: Methodology and empirical evidence for US banks. European Journal of Operational Research. 2020;**282**(2):786-801

[2] Carmona P, Climent F, Momparler A. Predicting failure in the U.S. banking sector: An extreme gradient boosting approach. International Review of Economics and Finance. 2019;**61**: 304-323

[3] Jing Z, Fang Y. Predicting US bank failures: A comparison of logit and data mining models. Journal of Forecasting. 2018;**37**:235-256

[4] Keramati A, Ghaneei H, Mirmohammadi SM. Developing a prediction model for customer churn from electronic banking services using data mining. Financial Innovation. 2016; **2**(1):1-13

[5] Lv F, Huang J, Wang W, Wei Y, Sun Y, Wang B. A two-route CNN model for bank account classification with heterogeneous data. PLoS One. 2019;**14**(8):1-22

[6] Wan J, Yue Z-L, Yang D-H, Zhang Y, Jiao L, Zhi L, et al. Predicting non performing loan of business Bank with data mining techniques. International Journal of Database Theory and Application. 2016;**9**(12):23-34

[7] Yang Q, Wu X. 10 challenging problems in data mining research. International Journal of Information Technology and Decision Making. 2006; **5**(4):597-604

[8] Marinakos G, Daskalaki S. Imbalanced customer classification for bank direct marketing. Journal of Marketing Analytics. 2017;**5**(1):14-30

[9] Smeureanu I, Ruxanda G, Badea LM. Customer segmentation in private banking sector using machine learning techniques. Journal of Business Economics and Management. 2013; **14**(5):923-939

[10] Ogwueleka FN, Misra S, Colomo-Palacios R, Fernandez L. Neural network and classification approach in identifying customer behavior in the banking sector: A case study of an international bank. Human Factors and Ergonomics in Manufacturing. 2015; **25**(1):28-42

[11] Ilham A, Khikmah L, Indra A, Ulumuddin A, Iswara I. Long-term deposits prediction: A comparative framework of classification model for predict the success of bank telemarketing. Journal of Physics Conference Series. 2019; **1175**(1):1-6

[12] Farooqi R, Iqbal N. Performance evaluation for competency of bank telemarketing prediction using data mining techniques. International Journal of Recent Technology and Engineering. 2019;**8**(2):5666-5674

[13] Lahmiri S. A two-step system for direct bank telemarketing outcome classification. Intelligent Systems in Accounting, Finance and Management. 2017;**24**(1):49-55

[14] Moro S, Cortez P, Rita P. A data-driven approach to predict the success of bank telemarketing. Decision Support Systems. 2014;**62**:22-31

[15] Krishna GJ, Ravi V, Reddy BV, Zaheeruddin M, Jaiswal H, Sai Ravi Teja P, et al. Sentiment classification of Indian Banks' Customer Complaints. In: Proceedings of IEEE Region 10 Annual International Conference. India; 17–20 October 2019. pp. 429-434

[16] Hassani H, Huang X, Silva E. Digitalisation and Big Data Mining in Banking. Big Data and Cognitive Computing. 2018;**2**(3):1-13

[17] Chawla NV, Bowyer KW, Hall LO, Kegelmeyer WP. SMOTE: Synthetic minority over-sampling technique. Journal of Artificial Intelligence Research. 2002;**16**:321-357

[18] Cieslak D, Liu W, Chawla S, Chawla N. A robust decision tree algorithms for imbalanced data sets. In: Proceedings of the Tenth SIAM International Conference on Data Mining (SDM 2010). Columbus, Ohio, USA; 29 Apr-1 May 2010. pp. 766-777

[19] Shotton J, Nowozin S, Sharp T, Winn J, Kohli P, Criminisi A. Decision jungles: Compact and rich models for classification. Advances in Neural Information Processing Systems. 2013; **26**:234-242

[20] Dua D, Graff C. UCI Machine Learning Repository. Irvine, CA: University of California, School of Information and Computer Science. 2019. Available from: http://archive.ics.uci.edu/ml

[21] Carcillo F, Borgne Y-A, Caelen O, Oble F, Bontempi G. Combining unsupervised and supervised learning in credit card fraud detection. Information Sciences. 2020 in press. DOI: 10.1016/j.ins.2019.05.042

[22] Yeh IC, Lien CH. The comparisons of data mining techniques for the predictive accuracy of probability of default of credit card clients. Expert Systems with Applications. 2009;**36**(2): 2473-2480

Chapter 5

Analytical Statistics Techniques of Classification and Regression in Machine Learning

Pramod Kumar, Sameer Ambekar, Manish Kumar and Subarna Roy

Abstract

This chapter aims to introduce the common methods and practices of statistical machine learning techniques. It contains the development of algorithms, applications of algorithms and also the ways by which they learn from the observed data by building models. In turn, these models can be used to predict. Although one assumes that machine learning and statistics are not quite related to each other, it is evident that machine learning and statistics go hand in hand. We observe how the methods used in statistics such as linear regression and classification are made use of in machine learning. We also take a look at the implementation techniques of classification and regression techniques. Although machine learning provides standard libraries to implement tons of algorithms, we take a look on how to tune the algorithms and what parameters of the algorithm or the features of the algorithm affect the performance of the algorithm based on the statistical methods.

Keywords: machine learning, statistics, classification, regression, algorithms

1. Introduction

Stating that statistical methods are useful in machine learning is analogous to saying that wood working methods are helpful for a carpenter. Statistics is the foundation of machine learning. However not all machine learning methods have been said to have derived from statistics. To begin with let us take a look at what statistics and machine learning means.

Statistics is extensively used in areas of science and finance and in the industry. Statistics is known to be mathematical science and not just mathematics. It is said to have been originated in seventeenth century. It consists of data collection, organizing the data, analyzing the data, interpretation and presentation of data. Statistical methods are being used since a long time in various fields to understand the data efficiently and to gain an in-depth analysis of the data [1].

On the other hand, machine learning is a branch of computer science which uses statistical abilities to learn from a particular dataset [2]. It was invented in the year 1959. It learns using algorithm and then has the ability to predict based on what it has been fed with. Machine learning gives out detailed information than statistics [3].

Most of the techniques of machine learning derive their behavior from statistics. However not many are familiar with this since both of them have their own jargons. For instance learning in statistics is called as fitting, supervised learning from machine learning is called as regression. Machine learning is a subfield of computer science and artificial intelligence. Machine learning is said to be a subdivision of computer science and artificial intelligence. It does use fewer assumptions than statistics. Machine learning unlike statistics deals with large amount of data and it also requires minimum human effort since most of its computation is done by the machine or the computer itself. Machine learning unlike statistics has a strong predicting power than statistics. Depending on the type of data machine learning can be categorized into supervised machine learning, unsupervised machine learning and reinforcement learning [4].

There seems to be analogy between machine learning and statistics. The following picture from textbook shows how statistics and machine learning visualize a model. **Table 1** shows how terms of statistics have been coined in machine learning.

To understand how machine learning and statistics come out with the results let's look at **Figure 1**. In statistical modeling on the left half of the image, linear regression with two variables is fitting the best plane with fewer errors. In machine learning the right half of the image to fit the model in the best possible way the independent variables have been converted into the square of error terms. That is machine learning strives to get a better fit than the statistical model. In doing so, machine learning minimizes the errors and increases the prediction rates.

Statistics methods are not just useful in training the machine learning model but they are helpful in many other stages of machine learning such as:

Machine learning	**Statistics**
Network, graphs	Model
Weights	Parameters
Learning	Fitting
Generalization	Tool set performance
Supervised learning	Regression/classification
Unsupervised learning	Density estimation, clustering

Table 1.
Machine learning jargons and corresponding statistics jargons.

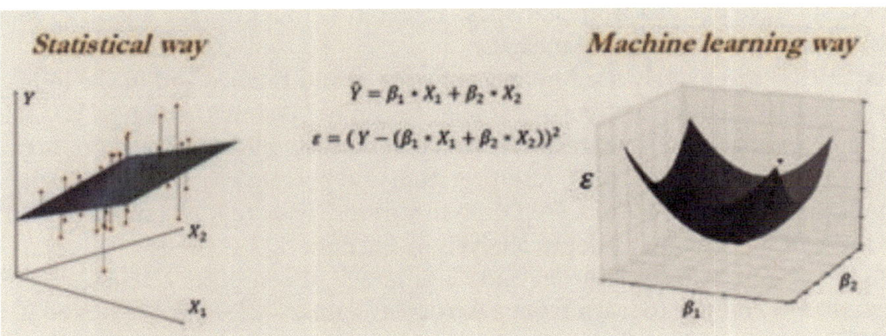

Figure 1.
Statistical and machine learning method.

- Data preparation—where statistics is used for data preprocessing which is later sent to the model. For instance when there are missing values in the dataset, we compute statistical mean or statistical median and fill it in the empty spaces of the dataset. It is recommended that machine learning model should never be fed with a dataset which has empty cells in it. It also used in preprocessing stage to scale the data by which the values are scaled to a particular range by which the mathematical computation becomes easy during the training of machine learning.

- Model evaluation—no model is perfect in predicting when it is built for the first time. Simply building the model is not enough. It is vital to check how well is it performing and if not then by how much is it closer to being accurate enough. Hence, we evaluate the model by statistical methods, which tell by how much the result is accurate and a lot many things about the end result obtained. We make use of metrics such as confusion matrix, Kolmogorov Smirnov chart, AUC—ROC, root mean squared error and many metrics to enhance our model.

- Model selection—we make use of many algorithms to train the algorithm and there is a chance of selecting only one which gives out accurate results when compared to others. The process of selecting the right solution for this is called model selection. Two of the statistical methods can be used to select the appropriate model such as statistical hypothesis test and estimation statistics [5].

- Data selection—some datasets carry a lot of features with them. Of many features, it may happen so that only some contribute significantly in estimation of the result. Considering all the features becomes computationally expensive and as well as time consuming. By making use of statistics concepts we can eliminate the features which do not contribute significantly in producing the result. That is it helps in finding out the dependent variables or features for any result. But it is important to note that this method requires careful and skilled approach. Without which it may lead to wrong results.

In this chapter we take a look at how statistical methods such as, regression and classification are used in machine learning with their own merits and demerits.

2. Regression

Regression is a statistical measure used in finance, investing and many other areas which aims to determine relationship between the dependent variables and 'n' number of independent variables. Regression consists of two types:

Linear regression—where one independent variable is used to explain or predict the outcome of the dependent variable.

Multiple regression—where two or more independent variables are used to explain or predict the outcome of the dependent variable.

In statistical modeling, regression analysis consists of set of statistical methods to estimate how the variables are related to each other.

Linear and logistic are the types of regression which are used in predictive modeling [6].

Linear assumes that the relationship between the variables are linear that is they are linearly dependent. The input variables consist of variables X_1, X_2, ..., X_n (where n is a natural number).

Linear models were developed long time ago but till date they are able to produce significant results. That is even in the modern computer's era they are well off. They are widely used because they are not complex in nature. In prediction, they can even out perform complex nonlinear models.

There are 'n' number of regressions that can be performed. We look at the most widely used five types of regression techniques. They are:

- Linear regression
- Logistic regression
- Polynomial regression
- Stepwise regression
- Ridge regression

Any regression method would involve the following:

- The unknown variables is denoted by beta
- The dependent variables also known as output variable
- The independent variables also known as input variables

It is denoted in the form of function as:

$$Y \approx f(X, \beta) \tag{1}$$

2.1 Linear regression

It is the most widely used regression type by far. Linear regression establishes a relationship between the input variables (independent variables) and the output variable (dependent variable).

$$\text{That is } Y = X_1 + X_2 + \dots + X_n$$

It assumes that the output variable is a combination of the input variables. A linear regression line is represented by Y = a + bX, where X is the explanatory variable and Y is the dependent variable. The slope of the line is 'b', and 'a' is the intercept (the value of y when x = 0).

A line regression is represented by the equation:

$$Y = a + bX$$

where X indicates independent variables and 'Y' is the dependent variable [7]. This equation when plotted on a graph is a line as shown below in **Figure 2**.

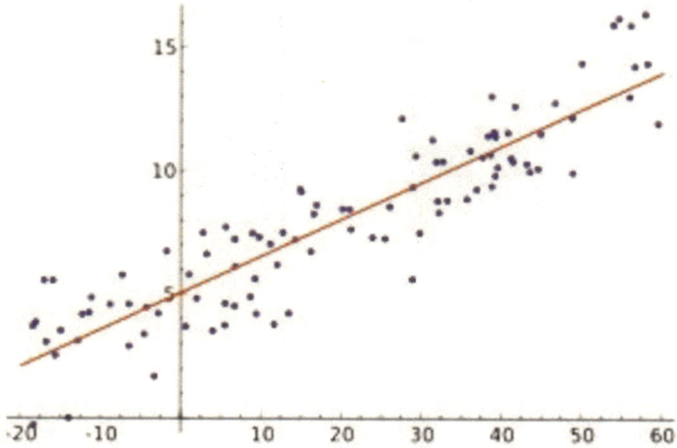

Figure 2.
Linear regression on a dataset.

However, linear regression makes the following assumptions:

- That there is a linear relationship
- There exists multivariate normality
- There exists no multi collinearity or little multicollinearity among the variables
- There exists no auto-correlation between the variables
- No presence of homoscedasticity

It is fast and easy to model and it is usually used when the relationship to be modeled is not complex. It is easy to understand. However linear regression is sensitive to outliners.

Note: In all of the usages stated in this chapter, we have assumed the following:

The dataset has been divided into training set (denoted by X) and test set (denoted by y_test)

The regression object "reg" has been created and exists.

We have used the following libraries:

Scipy and Numoy for numerical calculations

Pandas for dataset handling

Scikit-learn to implement the algorithm, to split the dataset and various other purposes.

Usage of linear regression in python:

```
import matplotlib.pyplot as plt
import numpy as np
from sklearn import datasets, linear_model
#Declare the linear regression function
reg=linear_model.LinearRegression()
#call the method
reg.fit(height,weight)
#to check slope and intercept
```

```
m=reg.coef_[0]
b=reg.intercept_
print("slope=",m, "intercept=",b)
# check the accuracy on the training set
reg.score(X, y)
```

2.2 Logistic regression

Logistic regression is used when the dependent variable is binary (True/False) in nature. Similarly the value of y ranges from 0 to 1 (**Figure 3**) and it is represented by the equation:

$$\text{Odds} = p/(1-p) = \text{probability that event will occur/probability that the event will not occur}$$

$$\ln(\text{odds}) = \ln(p/(1-p)) \tag{2}$$

$$\text{logit}(p) = \ln(p/(1-p)) = b_0 + b_1X_1 + b_2X_2 + b_3X_3 ... + b_kX_k$$

Logistic regression is used in classification problems. For example to classify emails as spam or not and to predict whether the tumor is malignant or not. It is not mandatory that the input variables have linear relationship to the output variable [8]. The reason being that it makes us of nonlinear log transformation to the predicted odds. It is advised to make use of only the variables which are powerful predictors to increase the algorithms performance.

However, it is important to note the following while making use of logistic regression:

Doesn't handle large number of categorical features.

The non-linear features should be transformed before using them.

Usage of logistic regression in python:

```
import numpy as np
import pandas as pd
from sklearn.linear_model import LogisticRegression
# instantiate a logistic regression model, and fit with X and y
reg = LogisticRegression()
reg = model.fit(X, y)
# check the accuracy on the training set
reg.score(X, y)
```

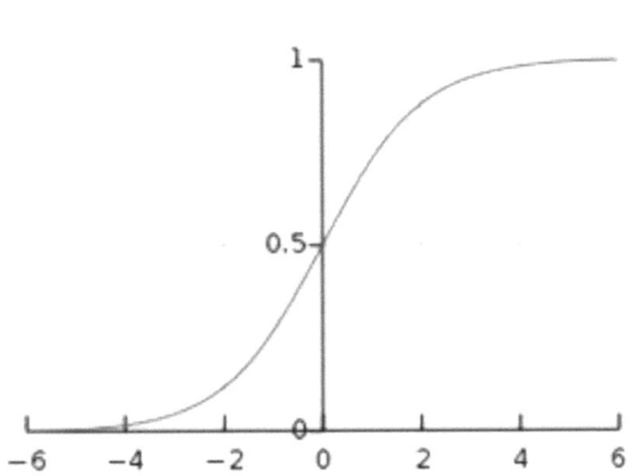

Figure 3.
Standard logistic function.

DOI: http://dx.doi.org/10.5772/intechopen.84922

2.3 Polynomial regression

It is a type of regression where the independent variable power is greater than 1. Example:

$$Y = a + b(X_2 + X_3 + ...X_n). \qquad (3)$$

The plotted graph is usually a curve in nature as shown in **Figure 4**.

If the degree of the equation is 2 then it is called quadratic. If 3 then it is called cubic and if it is 4 it is called quartic. Polynomial regressions are fit with the method of least squares. Since the least squares minimizes the variance of the unbiased estimators of all the coefficients which are done under the conditions of Gauss-Markov theorem. Although we may get tempted to fit a higher degree polynomial so that we could get a low error, it may cause over-fitting [9].

Some guidelines which are to be followed are:

The model is more accurate when it fed with large number of observations.

Not a good thing to extrapolate beyond the limits of the observed values.

Values for the predictor shouldn't be large else they will cause overflow with higher degree.

Usage of polynomial regression in python:

```
from sklearn.preprocessing import PolynomialFeatures
import numpy as np
#makes use of a pre-processor called degree for the function
reg = PolynomialFeatures(degree=2)
reg.fit_transform(X)
reg.score(X, y)
```

2.4 Step-wise regression

This type of regression is used when we have multiple independent variables. To select the variables which are independent an automatic process is used. If used in the right way it puts more power and presents us ton of information. It can be used when the number of variables is too many. However if it is used haphazardly it may affect the models performance.

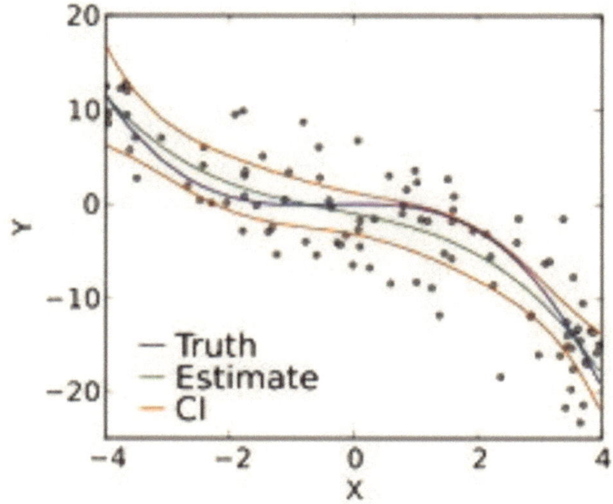

Figure 4.
Plotted graph is looks as curve in nature.

We make use of the following scores to help us find out the independent variables which contribute to the output variable significantly—R-squared, Adj. R-squared, F-statistic, Prob (F-statistic), Log-Likelihood, AIC, BIC and many more.

It can be performed by any of the following ways:

- Forward selection—where we start by adding the variables to the set and check how affects the scores.

- Backward selection—we start by taking all the variables to the set and start eliminating them one by one by looking at the score after each elimination.

- Bidirectional selection—a combination of both the methods mentioned above.

The greatest limitation of using step-wise regression is that the each instance or sample must have at least five attributes. Below which it has been observed that the algorithm doesn't perform well [10].

Code to implement Backward Elimination algorithm:

Assume that the dataset consists of 5 columns and 30 rows, which are present in the variable 'X' and let the expected results contain in the variable 'y'. Let 'X_opt' contain the independent variables which are used to determine the value of 'y'.

We are making use of a package called statsmodels, which is used to estimate the model and to perform statistical tests.

```
#import stats models package
import statsmodels.formula.api as sm
#since it is a polynomial add a column of 1s to the left
X = np.append (arr = np.ones([30,1]).astype(int), values = X, axis = 1)
#Let X-opt contain the independent variables only and Let y contain the output variable
X_opt = X[:,[0,1,2,3,4,5]]
#assign y to endog and X_opt to exog
regressor_OLS = sm.OLS(endog = y, exog = X_opt).fit()
regressor_OLS.summary()
```

The above code outputs the summary and based on it the variable which should be eliminated should be decided. Once decided remove the variable from 'X-opt'.

It is used to handle high dimensionality of the dataset.

2.5 Ridge regression

It can be used to analyze the data in detail. It is a technique which is used to get rid of multi collinearly. That is the independent values may be highly correlated. It adds a degree of bias due to which it reduces the standard errors.

The multi collinearity of the data can be inspected by correlation matrix. Higher the values, more the multi collinearity. It can also be used when number of predictor variables in the dataset exceeds the number of instances or observations [11].

The equation for linear regression is

$$Y = A + bX \tag{4}$$

This equation also contains error. That is it can be expressed as

$$Y = A + bX + (\text{error})$$

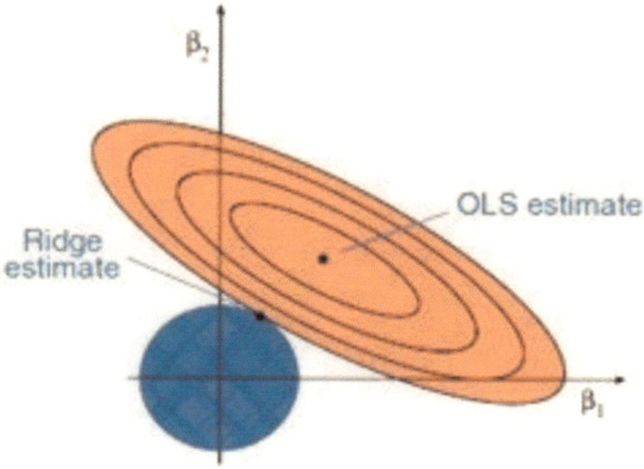

Figure 5.
Ridge and OLS.

Error with mean zero and known variance.

Ridge regression is known to shrink the size by imposing penalty on the size. It is also used to control the variance.

In (**Figure 5**) how ridge regression looks geometrically.

Usage of ridge regression in python:

```
from sklearn import linear_model
reg = linear_model.Ridge (alpha = .5)
reg.fit ([[0, 0], [0, 0], [1, 1]], [0, .1, 1])
Ridge(alpha=0.5, copy_X=True, fit_intercept=True, max_iter=None,
normalize=False, random_state=None, solver='auto', tol=0.001)
#to return the co-efficient and intercept
reg.coef_
reg.intercept_
```

2.6 Lasso regression

Least absolute shrinkage and selection operator is also known as LASSO. Lasso is a linear regression that makes use of shrinkage. It does so by shrinking the data values toward the mean or a central point. This is used when there are high levels of multi collinearity [12].

It is similar to ridge regression and in addition it can reduce the variability and improves the accuracy of linear regression models.

It is used for prostate cancer data analysis and other cancer data analysis.

Important points about LASSO regression:

- It helps in feature extraction by shrinking the co-efficient to zero.
- It makes use of L1 regularization.
- In the data if the predictors are have high correlation, the algorithm selects only one of the predictors discards the rest.

Code to implement in python:

```
from sklearn import linear_model
clf = linear_model.Lasso(alpha = 0.1)
```

```
clf.fit()
Lasso(alpha=0.1, copy_X=True, fit_intercept=True, max_iter=1000,
normalize=False, positive=False, precompute=False, random_state=None,
selection='cyclic', tol=0.0001, warm_start=False)
#to return the co-efficent and intercept
print(clf.coef_)
print(clf.intercept_)
```

3. Classification

A classification task is when the output is of the type "category" such as segregating data with respect to some property. In machine learning and statistics, classification consists of categorizing the new data to a particular category where it fits in on the basis of the data which has been used to train the model. Examples of tasks which make use of classification techniques are classifying emails as spam or not, detecting a disease on plants, predicting whether it will rain on some particular day, predicting the house prices based on the area it is located.

In terms of machine learning classification techniques fall under supervised learning [13].

The categories may be either:

- categorical (example: blood groups of humans—A, B, O)
- ordinal (example: high, medium or low)
- integer valued (example: occurrence of a letter in a sentence)
- Real valued

The algorithms which make use of this concept in machine learning and classify the new data are called as "Classifiers." Algorithms always return a probability score of belonging to the class of interest. That is considered an example where we are required to classify a gold ornament. Now when we input the image to the machine learning model the algorithms returns the probability value for each category, such as for if it is a ring the probability value may be higher than 0.8 if it not a necklace it may return less than 0.2, etc.

Higher the value more likely it is for it to belong to the particular group.

We make use of the following approach to build a machine learning classifier:

1. Pick a cut off probability above which we consider a record to belong to that class.

2. Estimate that a new observation belongs to a class.

3. If the obtained probability is above the cut off probability, assign the new observation to that class.

Classifiers are of two types: linear and nonlinear classifiers.

We now take a look at various classifiers are also statistical techniques:

1. Naive Bayes

2. stochastic gradient dissent (SGD)

3. K-nearest neighbors

4. decision trees

5. random forest

6. support vector machine

3.1 Naive Bayes

In machine learning, these classifiers belong to "probabilistic classifiers." This algorithm makes use of Bayes' theorem with strong independence assumptions between the features. Although Naive Bayes were introduced in the early 1950s, they are still being used today [14].

Given a problem instance to be classified, represented by a vector

$$X = (x_1, x_2, x_3, \ldots, x_n)$$

Which represent 'n' features.

$$P(Ck \mid x_1, x_2, \ldots, x_n)$$

We can observe that in the above formula that if the number of features is more or if a feature accommodates a large number of values, then it becomes infeasible. Therefore we rewrite the formula based on Bayes theorem as:

$$p(C_k|x) = p(C_k)p(x|C_k)/p(x) \tag{5}$$

Makes two "naïve" assumptions over attributes:

- All attributes are a priori equally important
- All attributes are statistically independent (value of one attribute is
- not related to a value of another attribute)

This classifier makes two assumptions:

- All attributes are equally important
- All attributes are not related to another attribute

There are three types of naive Bayes algorithms, which can be used: GaussianNB, BernoulliNB, and MultinomialNB.

Usage of naive Bayes in python:

```
import pandas as pd
import numpy as np
import matplotlib.pyplot as plt
from sklearn.naive_bayes import GaussianNB
reg= GaussianNB()
reg.fit(X,y)
reg.predict(X_test)
reg.score()
```

3.2 Stochastic gradient dissent (SGD)

An example of linear classifier which implements regularized linear model (**Figure 6**) with stochastic gradient dissent. Stochastic gradient descent (often shortened to SGD), also known as incremental gradient descent, is an iterative method to optimize a differentiable objective function, a stochastic approximation of gradient descent optimization [15]. Although SGD has been a part of machine learning since ages it wasn't extensively used until recently.

In linear regression algorithm, we make us of least squares to fit the line. To ensure that the error is low we use gradient descent. Although gradient descent does the job it can't handle big tasks hence we use stochastic gradient classifier. SGD calculates the derivative of each training data and also calculates the update within no time.

The advantages of using SGD classifier are that they are efficient and they are easy to implement.

However it is sensitive to feature scaling.

Usage of SGD classifier:

```
from sklearn.linear_model import SGDClassifier
X = [[0., 0.], [1., 1.]]
y = [0, 1]
clf = SGDClassifier (loss = "hinge", penalty = "l2")
clf.fit(X, y)
#to predict the values
clf.fit(X_test)
```

3.3 K-nearest neighbors

Also known as k-NN is a method used to classify as well as for regression. The input consists of k number of closest training examples. It is also referred as lazy learning since the training phase doesn't require a lot of effort.

In k-NN an object's classification is solely dependent on the majority vote of the object's neighbors. That is the outcome is based on the presence of the neighbors. The object is assigned to the class most common among its k nearest neighbors. If the value of k is equal to 1 then it's assigned to its nearest neighbor. Simply put, the

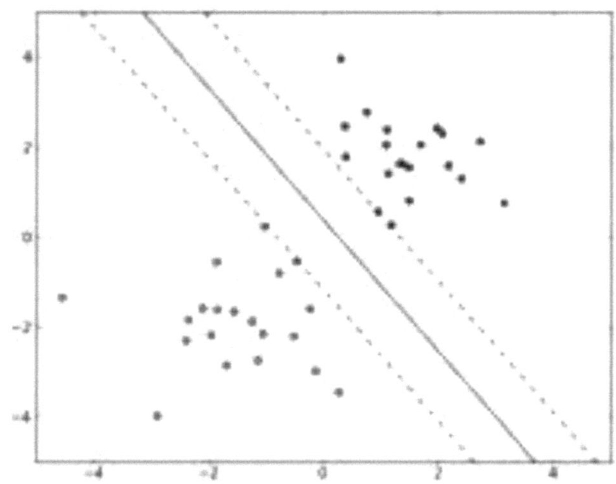

Figure 6.
Feature scaling classifier.

k-NN algorithm is entirely dependent on the neighbors of the object to be classified. Greater the influence of a neighbor, the object is assigned to it. It is termed as simplest machine learning algorithm among all the algorithms [16].

Let us consider an example where the green circle is the object which is to be classified as shown in **Figure 7**. Let us assume that there are two circles—the solid circle and the dotted circle.

As we know that there are two classes class 1 (blue squares) and class 2 (red squares). If we consider only the inner circle that is the solid circle then there are two objects of red circle existing which dominates the number of blue squares due to which the new object is classified to Class 1. But if we consider the dotted circle, the number of blue circle dominates since there are more number of blue squares due to which the object is classified to Class 2 [17].

However, the cost of learning process is zero.

The algorithm may suffer from curse of dimensionality since the number of dimensions greatly affects its performance. When the dataset is very large the computation becomes very complex since the algorithm takes time to look out for its neighbors. If there are many dimensions then the samples nearest neighbors can be far away. To avoid curse of dimensionality dimension reduction is usually performed before applying k-NN algorithm to the data.

Also the algorithm may not perform well with categorical data since it is difficult to find the distance between the categorical features.

Usage in python:

```
from sklearn.neighbors import KNeighborsClassifier
classifier = KNeighborsClassifier (n_neighbors=5)
classifier.fit(X_train, y_train)
```

3.4 Decision trees

Decision trees are considered to be most popular classification algorithms while classifying data. Decision trees are a type of supervised algorithm where the data is split based on certain parameters. The trees consist of decision nodes and leaves [18].

The decision tree consists of a root tree from where the tree generates and this root tree doesn't have any inputs. It is the point from which the tree originates. All the other nodes except the root node have exactly one incoming node. The other

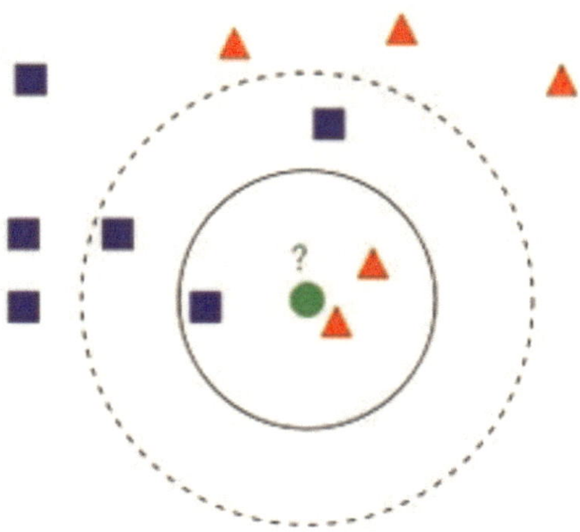

Figure 7.
K-Neighbors.

nodes except the root node are called leaves. Below is the example of a decision tree an illustration of how the decision tree looks like as shown in **Figure 8**.

"Is sex male" is the root node from where the tree originates. Depending on the condition the tree further bifurcates into subsequent leaf nodes. Few more conditions like "is Age >9.5?" are applied by which the depth of the node goes on increasing. As the number of leaf nodes increase the depth of the tree goes on increasing. The leaf can also hold a probability vector.

Decision tree algorithms implicitly construct a decision tree for any dataset.

The goal is to construct an optimal decision tree by minimalizing the generalization error. For any tree algorithm, it can be tuned by making changes to parameters such as "Depth of the tree," "Number of nodes," "Max features." However construction of a tree by the algorithm can get complex for large problems since the number of nodes increase as well as the depth of the tree increases.

Advantages of this tree are that they are simple to understand and can be easily interpreted. It also requires little data preparation. The tree can handle both numerical and categorical data unlike many other algorithms. It also easy to validate the decision tree model using statistical testes. However, disadvantages of the trees are that they can be complex in nature for some cases which won't generalize the data well. They are unstable in nature since if there are small variations in data they may change the structure of the tree completely.

Usage in python:

```
from sklearn.neighbors import tree
classifier = tree.DecisionTreeClassifier()
classifier.fit(X_train, y_train)
clf.predict(X_test)
```

3.5 Random forest

These are often referred as ensemble algorithms since these algorithms combine the use of two or more algorithms. They are improved version of bagged decision trees. They are used for classification, regression, etc.

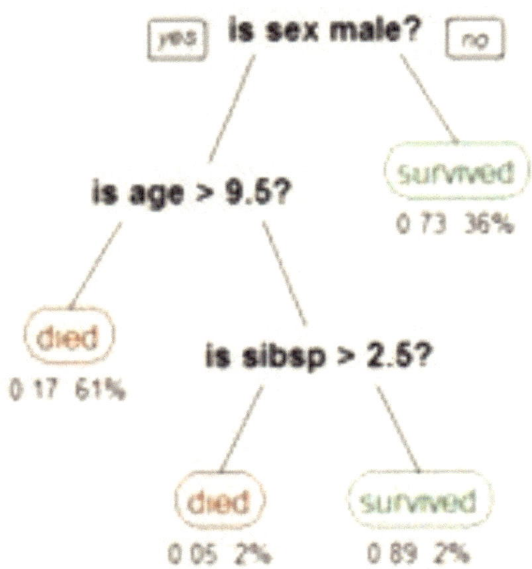

Figure 8.
Typical decision tree.

Random forest creates n number of decision trees from a subset of the data. On creating the trees it aggregates the votes from the different trees and then decides the final class of the sample object. Random forest is used in recommendation engines, image classification and feature selection [19].

The process consists of four steps:

1. It selects random samples from the dataset.

2. For every dataset construct a dataset and then predict from every decision tree.

3. For every predicted result perform vote.

4. Select the prediction which has the highest number of votes.

Random forest's default parameters often produce a good result in most of the cases. Additionally, one can make changes to achieve desired results. The parameters in Random Forest which can be used to tune the algorithm which can be used to give better and efficient results are:

1. Increasing the predictive power by increasing "n_estimators" by which the number of tress which will be built can be altered. "max_features" parameter can also be adjusted which is the number of features which are used to train the algorithm. Another parameter which can be adjusted is "min_sample_leaf" which is the number of leafs that are used to split the internal node.

2. To increase the model's speed, "n_jobs" parameter can be adjusted which is the number of processors it can use. To use as many as needed "−1" can be specified which signifies that there is no limit.

Due to large number of decision trees random forest is highly accurate. Since it takes the average of all the predictions which are computed the algorithm doesn't suffer from over fitting. Also it does handle missing values from the dataset. However, the algorithm is takes time to compute since it takes time to build trees and take the average of the predictions and so on.

One of the real time examples where random forest algorithm can be used is predicting a person's systolic blood pressure based on the person's height, age, weight, gender, etc.

Random forests require very little tuning when compared to other algorithms. The main disadvantage of random forest algorithm is that increased number of tress can make the process computationally expensive and lead to inaccurate results.

Usage in python:

```
from sklearn.ensemble import RandomForestClassifier
clf = RandomForestClassifier(max_depth=2, random_state=0)
clf.fit(X, y)
clf.predict(X_test)
```

3.6 Support vector machine

Support vector machines also known as SVMs or support vector networks fall under supervised learning. They are used for classification as well as regression purposes. Support vectors are the data points which lie close to the hyper plane. When the data is fed to the algorithm the algorithm builds a classifier which can be used to assign new examples to one class or the other [20]. A SVM consists of points in space separated by a gap which is as wide as possible. When a new sample is encountered it maps it to the corresponding category.

Perhaps when the data is unlabeled it becomes difficult for the supervised SVM to perform and this is where unsupervised method of classifying is required.

A SVM constructs a hyper plane which can be used for classification, regression and many other purposes. A good separation can be achieved when the hyper plane has the largest distance to the nearest training point of a class.

In (**Figure 9**) H_1 line doesn't separate, while H_2 separates but the margin is very small whereas H_3 separates such as the distance between the margin and the nearest point is maximum when compared to H_1 and H_2.

SVMs can be used in a variety of applications such as:

They are used to categorize text, to classify images, handwritten images can be recognized, and they are also used in the field of biology.

SVMs can be used with the following kernels:

1. Polynomial kernel SVM

2. Linear kernel SVM

3. Gaussian kernel SVM

4. Gaussian radial basis function SVM (RBF)

The advantages of SVM are:

1. Effective in high dimensional data

2. It is memory efficient

3. It is versatile

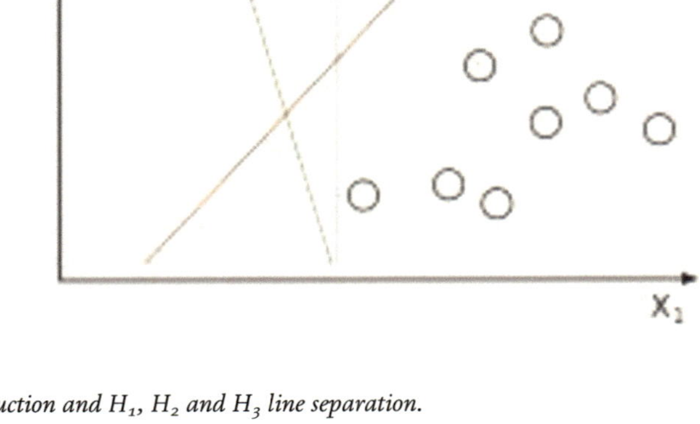

Figure 9.
Hyper plane construction and H_1, H_2 and H_3 line separation.

It may be difficult for SVM to classify at times due to which the decision boundary is not optimal. For example, when we want to plot the points randomly distributed on a number line.

It is almost impossible to separate them. So in such cases we transform the dataset by applying 2D or 3D transformations by using a polynomial function or any other appropriate function. By doing so it becomes easier to draw a hyper plane.

When the number of features is much greater than number of samples it doesn't perform well with the default parameters.

Usage of SVM in python:

```
from sklearn import svm
clf = svm.SVC()
clf.fit(X,y)
clf.predict(X_test)
```

4. Conclusion

It is evident from the above regression and classification techniques are strongly influenced by statistics. The methods have been derived from statistical methods which existed since a long time. Statistical methods also consist of building models which consists of parameters and then fitting it. However not all the methods which are being used derive their nature from statistics. Not all statistical methods are being used in machine learning. Extensive research in the field of statistical methods may give out new set methods which can be used in machine learning apart from the existing statistical methods which are being used today. It can also be stated that machine learning to some extent is a form of 'Applied Statistics.'

Author details

Pramod Kumar[1*], Sameer Ambekar[1†], Manish Kumar[2] and Subarna Roy[1]

1 Department of Health Research, Biomedical Informatics Centre, ICMR-National Institute of Traditional Medicine, Belagavi, Karnataka, India

2 Department of Electrical Engineering, College of Engineering, Bharti Vidyapeeth, Pune, Maharashtra, India

*Address all correspondence to: pramodbiotech@gmail.com

† Sameer Ambekar shares first authorship.

References

[1] Hawkins DM. On the investigation of alternative regressions by principal component analysis. Journal of the Royal Statistical Society Series. 1973;**22**: 275-286. https://www.jstor.org/stable/i316057

[2] Kourou K, Exarchos TP, Exarchos KP, Karamouzis MV, Fotiadis DI. Machine learning applications in cancer prognosis and prediction. Computational and Structural Biotechnology Journal. 2015;**13**:8-17. DOI: 10.1016/j.csbj.2014.11.005

[3] Machine Learning [Internet]. Available from: https://en.wikipedia.org/wiki/Machine_learning

[4] Trevor H, Robert T. The Elements of Statistical Learning: Data Mining, Inference, and Prediction. New York: Springer; 2009. pp. 485-586. DOI: 10.1007/978-0-387-84858-7_14

[5] Aho K, Derryberry DW, Peterson T. Model selection for ecologists: The worldviews of AIC and BIC. Ecology. 2014;**95**(3):631-636. DOI: 10.1890/13-1452.1

[6] Freedman DA. Statistical Models: Theory and Practice. USA: Cambridge University Press; 2005. ISBN: 978-0-521-85483-2

[7] sklearn.linear_model. LinearRegression—scikit-learn 0.19.2 documentation [Internet]. Available from: http://scikit-learn.org/stable/modules/generated/sklearn.linear_model.LinearRegression.html

[8] Linear Regression—Wikipedia [Internet]. Available from: https://en.wikipedia.org/wiki/Linear_regression

[9] Shaw P et al. Gergonne's 1815 paper on the design and analysis of polynomial regression experiments. Historia Mathematica; 2006;**1**(4):431-439. DOI: 10.1016/0315-0860(74)90033-0

[10] Stepwise Regression—Wikipedia [Internet]. Available from: https://en.wikipedia.org/wiki/Stepwise_regression

[11] Tikhonov Regularization—Wikipedia [Internet]. Available from: https://en.wikipedia.org/wiki/Tikhonov_regularization

[12] sklearn.linear_model. LogisticRegression—scikit-learn 0.19.2 documentation [Internet]. Available from: http://scikit-learn.org/stable/modules/generated/sklearn.linear_model.LogisticRegression.html

[13] Statistical Classification—Wikipedia [Internet]. Available from: https://en.wikipedia.org/wiki/Statistical_classification

[14] Naive Bayes Scikit-Learn 0.19.2 Documentation [Internet]. Available from: http://scikit-learn.org/stable/modules/naive_bayes.html

[15] Stochastic Gradient Descent—Scikit-Learn 0.19.2 Documentation [Internet]. Available from: http://scikit-learn.org/stable/modules/sgd.html

[16] k-Nearest Neighbors Algorithm—Wikipedia [Internet]. Available from: https://en.wikipedia.org/wiki/K-nearest_neighbors_algorithm

[17] Kamiński B, Jakubczyk M, Szufel P. A framework for sensitivity analysis of decision trees. Central European Journal of Operations Research. 2017;**26**: 135-159. DOI: 10.1007/s10100-017-0479-6

[18] Lasso (Statistics)—Wikipedia [Internet]. Available from: https://en.wikipedia.org/wiki/Lasso_(statistics)

[19] sklearn.linear_model.Lasso—Scikit-Learn 0.19.2 Documentation [Internet]. Available from: http://scikit-learn.org/stable/modules/generated/sklearn.linear_model.Lasso.html

[20] Corinna C, Vapnik Vladimir N. Support-vector networks. Machine Learning. 1995;**20**(3):273-297. DOI: 10.1007/BF00994018

Chapter 6

Clustering of Time-Series Data

Esma Ergüner Özkoç

Abstract

The process of separating groups according to similarities of data is called "clustering." There are two basic principles: (i) the similarity is the highest within a cluster and (ii) similarity between the clusters is the least. Time-series data are unlabeled data obtained from different periods of a process or from more than one process. These data can be gathered from many different areas that include engineering, science, business, finance, health care, government, and so on. Given the unlabeled time-series data, it usually results in the grouping of the series with similar characteristics. Time-series clustering methods are examined in three main sections: data representation, similarity measure, and clustering algorithm. The scope of this chapter includes the taxonomy of time-series data clustering and the clustering of gene expression data as a case study.

Keywords: time-series data, data mining, data representation, similarity measure, clustering algorithms, gene expression data clustering

1. Introduction

The rapid development of technology has led to the registration of many processes in an electronic environment, the storage of these records, and the accessibility of these records when requested. With the evolving technology such as cloud computing, big data, the accumulation of a large amount of data stored in databases, and the process of parsing and screening useful information made data mining necessary.

It is possible to examine the data which are kept in databases and reach to huge amounts of size every second, in two parts according to their changes in time: static and temporal. Data is called the static data when its feature values do not change with time, if the feature comprise values change with time then it is called the temporal or time-series data.

Today, with the increase in processor speed and the development of storage technologies, real-world applications can easily record changing data over time.

Time-series analysis is a trend study subject because of its prevalence in various fields ranging from science, engineering, bioinformatics, finance, and government to health-care applications [1–3]. Data analysts are looking for the answers of such questions: Why does the data change this way? Are there any patterns? Which series show similar patterns? etc. Subsequence matching, indexing, anomaly detection, motif discovery, and clustering of the data are the answers of some questions [4]. Clustering, which is one of the most important concepts of data mining, defines its structure by separating unlabeled data sets into homogeneous groups. Many general-purpose clustering algorithms are used for the clustering of time-series data, either by directly or by evolving. Algorithm selection depends entirely on the purpose of the application and on the properties of the data such as sales data, exchange rates in finance, gene expression data, image data for face recognition, etc.

In the age of informatics, the analysis of multidimensional data that has emerged as part of the digital transformation in every field has gained considerable importance. These data can be from data received at different times from one or more sensors, stock data, or call records to a call center. This type of data, that is, observing the movement of a variable over time, where the results of the observation are distributed according to time, is called time-series data. Time-series analysis is used for many purposes such as future forecasts, anomaly detection, subsequence matching, clustering, motif discovery, indexing, etc. Within the scope of this study, the methods developed for the time-series data clustering which are important for every field of digital life in three main sections. In the first section, the proposed methods for the preparation of multidimensional data for clustering (dimension reduction) in the literature are categorized. In the second section, the similarity criteria to be used when deciding on the objects to be assigned to the related cluster are classified. In the third section, clustering algorithms of time-series data are examined under five main headings according to the method used. In the last part of the study, the use of time-series clustering in bioinformatics which is one of the favorite areas is included.

2. Time-series clustering approaches

There are many different categorizations of time-series clustering approaches. Such as, time-series clustering approaches can be examined in three main sections according to the characteristics of the data used whether they process directly on raw data, indirectly with features extracted from the raw data, or indirectly with models built from the raw data [5]. Another category is according to the clustering method: shape-based, feature-based, and model-based [6]. But whatever the categorization is, for any time-series clustering approach, the main points to be considered are: how to measure the similarity between time series; how to compress the series or reduce dimension and what algorithm to use for cluster. Therefore, this chapter examines time-series clustering approaches according to three main building blocks: data representation methods, distance measurements, and clustering algorithms (**Figure 1**).

2.1 Data representation

Data representation is one of the main challenging issues for time-series clustering. Because, time-series data are much larger than memory size [7, 8] that increases the need for high processor power and time for the clustering process increases exponentially. In addition, the time-series data are multidimensional, which is a difficulty for many clustering algorithms to handle, and it slows down the calculation of the similarity measurement. Consequently, it is very important for time-series data to represent the data without slowing down the algorithm execution time and without a significant data loss. Therefore, some requirements can be listed for any data representation methods [9]:

i. Significantly reduce the data size/dimensionality,

ii. Maintain the local and global shape characteristics of the time series,

iii. Acceptable computational cost,

iv. Reasonable level of reconstruction from the reduced representation,

v. Insensitivity to noise or implicit noise handling.

Dimension reduction is one of the most frequently used methods in the literature [7, 10–12] for the data representation.

Definition:

The representation of a time series T with length n is a model $\bar{T}$ with reduced dimensions, so that T approximates T [13]. Dimension reduction or feature extraction is a very useful method for reducing the number of variables/attributes or units

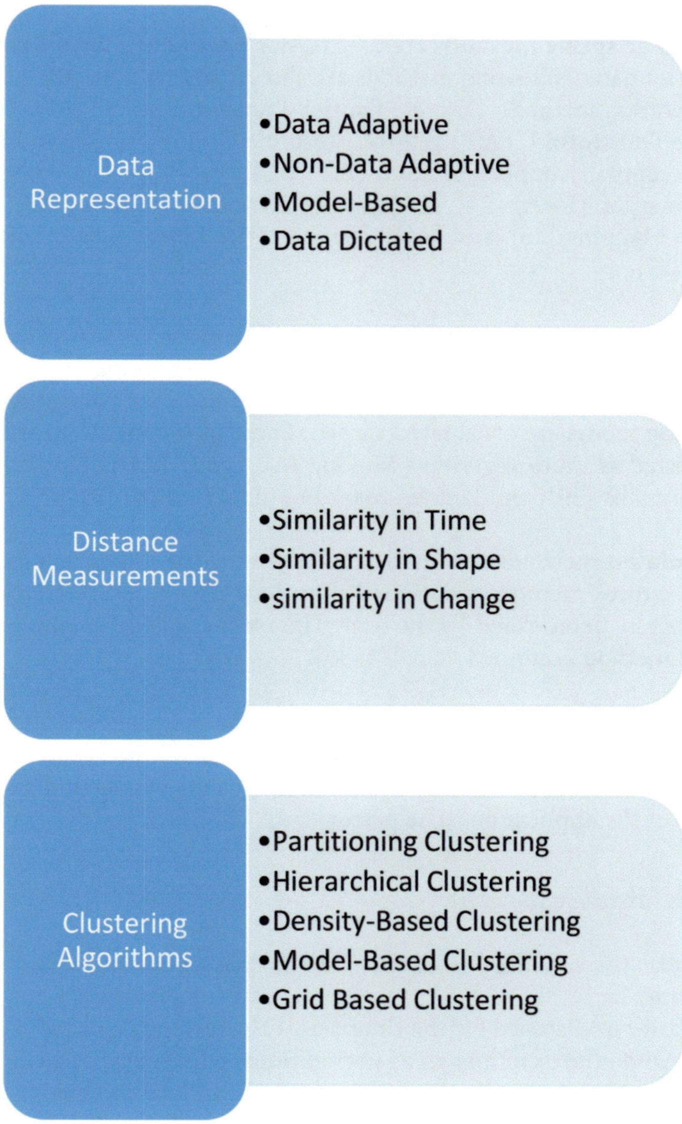

Figure 1.
Time-series clustering.

in multivariate statistical analyzes so that the number of attributes can be reduced to a number that "can handle."

Due to the noisy and high-dimensional features of many time-series data, data representations have been studied and generally examined in four main sections: data adaptive, nondata adaptive, model-based, and data dictated [6].

- **Data adaptive methods** that have changing parameters according to processing time-series data. Methods in this category try to minimize global reconstruction error by using unequal length segments. Although it is difficult to compare several time series, this method approximates each series better. Some of the popular data adaptive representation methods are: Symbolic Aggregate Approximation (SAX) [14], Adaptive Piecewise Constant Approximation (APCA) [15], Piecewise Linear Approximation (PLA) [16], Singular Value Decomposition (SVD) [17, 18], and Symbolic Natural Language (NLG) [19].

- **Non-data adaptive methods** are use fix-size parameters for the representing time-series data. Following methods are shown among non-data adaptive representation methods: Discrete Fourier Transform (DFT) [18], Discrete Wavelet Transform (DWT) [20–22], Discrete Cosine Transformation (DCT) [17], Perceptually Important Point (PIP) [23], Piecewise Aggregate Approximation (PAA) [24], Chebyshev Polynomials (CHEB) [25], Random Mapping [26], and Indexable Piecewise Linear Approximation (IPLA) [27].

- **Model-based methods** assume that observed time series was produced by an underlying model. The real issue here is to find the parameters that produce this model. Two time series produced by the same set of parameters using the underlying model are considered similar. Some of the model-based methods can be listed as: Auto-regressive Moving Average (ARMA) [28, 29], Time-Series Bitmaps [30], and Hidden Markov Model (HMM) [31–33].

- **Data dictated methods** automatically determine the dimension reduction rate but in the three methods mentioned above, the dimension reduction rates are automatically determined by the user. The most common example of data dictated method is clipped data [34–36].

Many representation methods for time-series data are proposed and each of them offering different trade-offs between the aforementioned requirements. The correct selection of the representation method plays a major role in the effectiveness and usability of the application to be performed.

2.2 Similarity/distance measure

In particular, the similarity measure is the most essential ingredient of time-series clustering.

The similarity or distance for the time-series clustering is approximately calculated, not based on the exact match as in traditional clustering methods. It requires to use distance function to compare two time series. In other words, the similarity of the time series is not calculated, it is estimated. If the estimated distance is large, the similarity between the time series is less and vice versa.

Definition:

Similarity between two "n" sized time series $T = \{t_1,t_2,....t_n\}$ and $U = \{u_1,u_2,....u_n\}$ is the length of the path connecting pair of points [11]. This distance is the measure of similarity. D (T, U) is a function that takes two times series (T, U) as input and calculates their distance "d".

Metrics to be used in clustering must cope with the problems caused by common features of time-series data such as noise, temporal drift, longitudinal scaling, offset translation, linear drift, discontinuities, and amplitude scaling. Various methods have been developed for similarity measure, and the method to choose is problem specific. These methods can be grouped under three main headings: similarity in time, similarity in shape, and similarity in change.

2.2.1 Similarity in time

The similarity between the series is that they are highly time dependent. Such a measure is costly for the raw time series, so a preprocessing or transformation is required beforehand [34, 36].

2.2.2 Similarity in shape

Clustering algorithms that use similarity in shape measure, assigns time series containing similar patterns to the same cluster. Independently of the time, it does not care how many times the pattern exists [37, 38].

2.2.3 Similarity in change

The result of using this metric is time-series clusters that have the similar autocorrelation structure. Besides, it is not a suitable metric for short time series [29, 39, 40].

2.3 Clustering algorithms

The process of separating groups according to similarities of data is called "clustering." There are two basic principles: the similarity within the cluster is the highest and the similarity between the clusters is the least. Clustering is done on the basis of the characteristics of the data and using multivariate statistical methods. When dividing data into clusters, the similarities/distances of the data to each other are measured according to the specification of the data (discrete, continuous, nominal, ordinal, etc.)

Han and Kamber [41] classify the general-purpose clustering algorithms which are actually designed for static data in five main sections: partition-based, hierarchical-based, density-based, grid-based, and model-based. Besides these, a wide variety of algorithms has been developed for time-series data. However, some of these algorithms (ignore minor differences) intend to directly use the methods developed for static data without changing the algorithm by transforming it into a static data form from temporal data. Some approaches apply a preprocessing step on the data to be clustered before using the clustering algorithm. This preprocessing step converts the raw-time-series data into feature vectors using dimension reduction techniques, or converts them into parameters of a specified model [42].

Definition:

Given a dataset on n time series $T = \{t_1, t_2, \ldots, t_n\}$, time-series clustering is the process of partitioning of T into $C = \{C_1, C_2, \ldots, C_k\}$ according to certain similarity criterion. C_i is called "cluster" where,

$$T = \bigcup_{i=1}^{k} C_i \text{ and } C_i \bigcap C_j = \emptyset \text{ for } i \neq j \quad (1)$$

In this section, previously developed clustering algorithms will be categorized. Some of these algorithms work directly with raw time-series data, while others use the data presentation techniques that are previously mentioned.

Clustering algorithms are generally classified as: partitioning, hierarchical, graph-based, model-based, and density-based clustering.

2.3.1 Partitioning clustering

The K-means [43] algorithm is a typical partition-based clustering algorithm such that the data are divided into a number of predefined sets by optimizing the predefined criteria. The most important advantage is its simplicity and speed. So it can be applied to large data sets. However, the algorithm may not produce the same result in each run and cannot handle the outlier. Self-organizing map [44] is stronger than the noisy data clustering from K-means. The user is prompted to enter the cluster number and grid sets. It is difficult to determine the number of clusters for time-series data. Other examples of partition-based clustering are CLARANS [45] and K-medoids [46]. In addition, the partitioning approach is suitable for low-dimensional, well-separated data. However, time-series data are multidimensional and often contain intersections, embedded clusters.

In essence, these algorithms act as n-dimensional vectors to time-series data and applies distance or correlation functions to determine the amount of similarity between two series. Euclidean distance, Manhattan distance, and Pearson correlation coefficient are the most commonly used functions.

2.3.2 Hierarchical clustering

Contrary to the partitioning approach, which aims segmenting data that do not intersect, the hierarchical approach produces a hierarchical series of nested clusters that can be represented graphically (dendrogram, tree-like diagram). The branches of the dendrogram show the similarity between the clusters as well as the knowledge of the shaping of the clusters. Determined number of clusters can be obtained by cutting the dendrogram at a certain level.

Hierarchical clustering methods [47–49] are based on the separating clusters into subgroups that are processed step by step as a whole, or the stepwise integration of individual clusters into a cluster [50]. Hierarchical clustering methods are divided into two methods: agglomerative clustering methods and divisive hierarchical clustering methods according to the creation of the dendrogram.

In agglomerative hierarchical clustering methods, each observation is initially treated as an independent cluster, and then repeatedly, until each individual observation obtains a single set of all observations, thereby forming a cluster with the closest observation.

In the divisive hierarchical clustering methods, initially all observations are evaluated as a single cluster and then repeatedly separated in such a way that each observation is separated from the farthest observation to form a new cluster. This process continues until all the observations create a single cluster.

Hierarchical clustering not only forms a group of similar series but also provides a graphical representation of the data. Graphical presentation allows the user to have an overall view of the data and an idea of data distribution. However, a small change in the data set leads to large changes in the hierarchical dendrogram. Another drawback is high computational complexity.

2.3.3 Density-based clustering

The density-based clustering approach is based on the concepts of density and attraction of objects. The idea is to create clusters of dense multi-dimensional areas where objects attract each other. In the core of dense areas, objects are very close together and crowded. The objects in the walls of the clusters were scattered less frequently than the core. In other words, density-based clustering determines dense areas of object space. The clusters are dense areas which are separated by rare dense areas. DBSCAN [51] and OPTICS [52] algorithms are the most known of density-based clustering examples.

The density-based approach is robust for noisy environments. The method also deals with outliers when defining embedded clusters. However, density-based clustering techniques cause difficulties due to high computational complexity and input parameter dependency when the dimensional index structure is not used.

2.3.4 Model-based clustering

The model-based approach [53–55] uses a statistical infrastructure to model the cluster structure of the time-series data. It is assumed that the underlying probability distributions of the data come from the final mixture. Model-based algorithms usually try to estimate the likelihood of the model parameters by applying some statistical techniques such as Expectation Maximization (EM). The EM algorithm iterates between an "E-step," which computes a matrix z such that z_{ik} is an estimate of the conditional probability that observation i belongs to group k given the current parameter estimates, and an "M-step," which computes maximum likelihood parameter estimates given z. Each data object is assigned to a cluster with the highest probability until the EM algorithm converges, so as to maximize likelihood for the entirety of the grant.

The most important advantage of the model-based approach is to estimate the probability that i. observation belongs to k. cluster. In some cases, the time series is likely to belong to more than one cluster. For such time-series data, the probability-giving function of the approach is the reason for preference. In this approach, it is assumed that the data set has a certain distribution but this assumption is not always correct.

2.3.5 Grid-based clustering

In this approach, grids made up of square cells are used to examine the data space. It is independent of the number of objects in the database due to the used grid structure. The most typical example is STING [56], which uses various levels of quadrilateral cells at different levels of resolution. It precalculates and records statistical information about the properties of each cell. The query process usually begins with a high-level hierarchical structure. For each cell at the current level, the confidence interval, which reflects the cell's query relation, is computed. Unrelated cells are exempt from the next steps. The query process continues for the corresponding cells in the lower level until reaching the lowest layer.

After analyzing the data set and obtaining the clustering solution, there is no guarantee of the significance and reliability of the results. The data will be clustered even if there is no natural grouping. Therefore, whether the clustering solution obtained is different from the random solution should be determined by applying some tests. Some methods developed to test the quality of clustering solutions are classified into two types: external index and internal index.

- The external index is the most commonly used clustering evaluation method also known as external validation, external criterion. The ground truth is the goal clusters, usually created by experts. This index measures how well the target clusters and the resulting clusters overlap. Entropy, Adjusted Rand Index (ARI), F-measure, Jaccard Score, Fowlkes and Mallows Index (FM), and Cluster Similarity Measure (CSM) are the most known external indexes.

- The internal indexes evaluate clustering results using the features of data sets and meta-data without any external information. These are often used in cases where the correct solutions are not known. Sum of squared error is one of the most used internal methods which the distance to the nearest cluster determines the error. So clusters with similar time series are expected to give lower error values. Distance between two clusters (CD) index, root-mean-square standard deviation (RMSSTD), Silhouette index, R-squared index, Hubert-Levin index, semi-partial R-squared (SPR) index, weighted inter-intra index, homogeneity index, and separation index are the common internal indexes.

2.3.6 Clustering algorithm example: FunFEM

The funFEM algorithm [55, 57] allows to cluster time series or, more generally, functional data. FunFem is based on a discriminative functional mixture model (DFM) which allows the clustering of the curves (data) in a functional subspace. If the observed curves are $\{x_1, x_2...x_n\}$, FunFem aims cluster into K homogenous groups. It assumes that there exists an unobserved random variable $Z = \{z_1, z_2...z_n\} \in \{0, 1\}^k$, if x belongs to group k, Z_k is defined as 1 otherwise 0. The clustering task goal is to predict the value $z_i = (z_{i1},... z_{ik})$ of Z for each observed curve x_i, for i = 1...n. The FunFem algorithm alternates, over the three steps of Fisher EM algorithm [57] ("F-step," "E-Step" and "M-step") to decide group memberships of $Z = \{z_1, z_2...z_n\}$. In other words, from 12 defined discriminative functional mixture (DFM) models, Fisher-EM decides which data fit the best. The Fisher-EM algorithm alternates between three steps:

- an E step in which posterior probabilities that observations belong to the K groups are computed,

- an F step that estimates the orientation matrix U of the discriminative latent space conditionally to the posterior probabilities,

- an M step in which parameters of the mixture model are estimated in the latent subspace by maximizing the conditional expectation of the complete likelihood.

Fisher-EM algorithm updates the parameters repeatedly until the Aitken criterion is provided. Aitken criterion estimates the asymptotic maximum of the log-likelihood in order to detect in advance the algorithm converge [57]. In model-based clustering, a model is defined by its number of component/cluster K and its parameterization. In model selection task, several models are reviewed while selecting the most appropriate model for the considered data.

FunFEM allows to choose between AIC (Akaike Information Criterion) [58], BIC (Bayesian information criteria) [59], and ICL (Integrated Completed Likelihood) [60] when deciding the number of clusters. The penalty terms are: $\frac{\gamma(M)}{2}\log(n)$ in the BIC criterion, $\gamma(M)$ in the AIC criterion, and $\sum_{i=1}^{n}\sum_{k=1}^{K} t_{ik}\log(t_{ik})$ in the ICL criterion. Here, M indicates the number of parameters in the model, n is the number of observations, K is the number of clusters, and t_{ik} is the probability of ith observation belonging to kth cluster.

FunFem is implemented in R programming languages and serves as a function [61]. The algorithm is applied on a time series gene expression data in the following section. Input of the algorithm is gene expression data which is given in **Table 1**. The table shows the gene expression values measured as a result of the microarray experiment. The measurement was performed at six different times for each gene. The data were taken from the GEO database (GSE2241) [62]. FunFEM method is decided, and the best model is DkBk with $K = 4$ (bic = −152654.5) for input data. As a result, method assigned each gene to the appropriate cluster which is determined by the algorithm. **Table 2** demonstrates the gene symbol and cluster number. As a result, method assigned each gene to the appropriate cluster which is determined by the algorithm (**Table 2**).

Gene Symbol	TP1	TP2	TP3	TP4	TP5	TP6
AADAC	18.4	29.7	30	79.7	86.7	163.2
AAK1	253.2	141.8	49.2	118.7	145.2	126.7
AAMP	490	340.9	109.1	198.4	210.5	212
AANAT	5.6	1.4	3.7	3.1	1.6	4.9
AARS	1770	793.6	226.5	1008.9	713.3	1253.7
AASDHPPT	940.1	570.5	167.2	268.6	683	263.5
AASS	10.9	1.9	1.5	4.1	19.7	25.5
AATF	543.4	520.1	114.5	305.7	354.2	384.9
AATK	124.5	74.5	17	25.6	64.6	13.6
.	.	.	.	.	.	.
.	.	.	.	.	.	.
ZP2	4.1	1.4	0.8	1.4	1.4	3
ZPBP	23.4	13.7	7	7.8	22.3	26.9
ZW10	517.1	374.5	72.6	240.8	345.7	333.1
ZWINT	1245.4	983.4	495.3	597.4	1074.3	620.7
ZYX	721.6	554.9	135.5	631.5	330.9	706.8
ZZEF1	90.5	49.3	18.6	66.7	10.4	52.2
ZZZ3	457.3	317.1	93	243.2	657.5	443

Table 1.
Input data of the FunFEM algorithm.

Gene symbol	Cluster number
AADAC	2
AAK1	3
AAMP	3
AANAT	1
AARS	4
AASDHPPT	3
AASS	1
AATF	3
AATK	2
.	.
.	.
ZP2	1
ZPBP	1
ZW10	3
ZWINT	4
ZYX	4
ZZEF1	2
ZZZ3	3

Table 2.
Output data of the FunFEM algorithm.

3. Clustering approaches for gene expression data clustering

The approach to be taken depends on the application area and the characteristics of the data. For this reason, as a case study, the clustering of gene expression data, which is a special area of clustering of time-series data, will be examined in this section. Microarray is the technology which measures the expression levels of large numbers of genes simultaneously. DNA microarray technology overcomes traditional approaches in the identification of gene copies in a genome, in the identification of nucleotide polymorphisms and mutations, and in the discovery and development of new drugs. It is used as a diagnostic tool for diseases. DNA microarrays are widely used to classify gene expression changes in cancer cells.

The gene expression time series (gene profile) is a set of data generated by measuring expression levels at different cases/times in a single sample. Gene expression time series have two main characteristics, short and unevenly sampled. In The Stanford Microarray database, more than 80% of the time-series experiments contains less than 9 time points [63]. Observations below 50 are considered to be quite short for statistical analysis. Gene expression time-series data are separated from other time-series data by this characteristics (business, finance, etc.). In addition to these characteristics, three basic similarity requirements can be identified for the gene expression time series: scaling and shifting, unevenly distributed sampling points, and shape (internal structure) [64]. Scaling and shifting problems arise due to two reasons: (i) the expression of genes with a common sequence is similar, but in this case, the genes need not have the same level of expression at the same time. (ii) Microarray technology, which is often corrected by normalization.

The scaling and shifting factor in the expression level may hide similar expressions and should not be taken into account when measuring the similarity between the two expression profiles. Sampling interval length is informative and cannot be ignored in similarity comparisons. In microarray experiments, the density change characterizes the shape of the expression profile rather than the density of the gene expression. The internal structure can be represented by deterministic function, symbols describing the series, or statistical models.

There are many popular clustering techniques for gene expression data. The common goal of all is to explain the different functional roles of the genes that play a key biological process. Genes expressed in a similar way may have a similar functional role in the process [65].

In addition to all these approaches, it is possible to examine the cluster of gene expression data in three different classes as gene-based clustering, sample-based clustering, and subspace clustering (**Figure 2**) [66]. In gene-based clustering, genes are treated as objects, instances (time-point/patient-intact) as features. Sample-based clustering is exactly the opposite: samples are treated as objects, genes as features. The distinction between these two clustering approaches is based on the basic characterization of the clustering process used for gene expression data. Some clustering algorithms, such as K-means and hierarchical approach, can be used to cluster both genes and fragments of samples. In the molecular biology, "any function in the cell is carried out with the participation of a small subset of genes, and the cellular function only occurs on a small sample subset." With this idea, genes and samples are handled symmetrically in subspace clustering; gene or sample, object or features.

In **gene-based clustering**, the aim is to group the co-expressed genes together. However, due to the complex nature of microarray experiments, gene expression data often contain high amounts of noise, characterizing features such as gene expression data often linked to each other (clusters often have a high intersection ratio), and some problems arising from constraints from the biological domain.

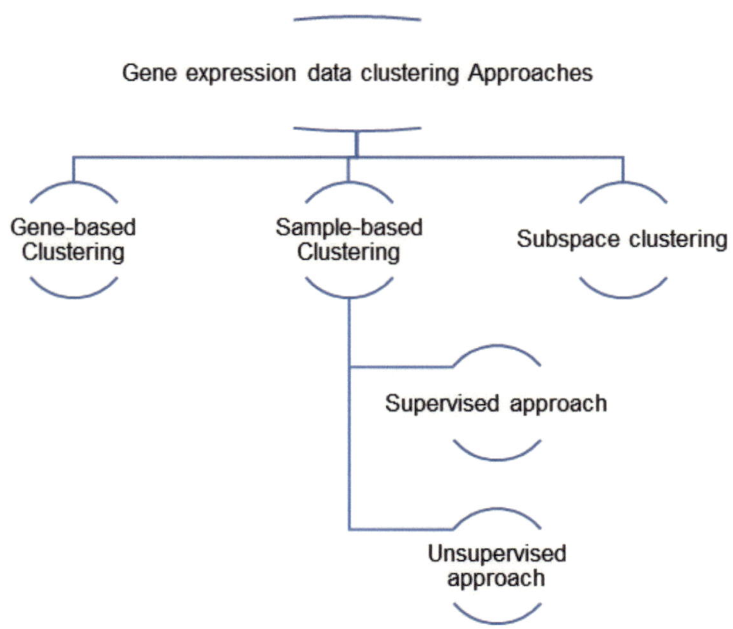

Figure 2.
Gene expression data clustering approaches.

Also, among biologists who will use microarray data, the relationship between genes or clusters that are usually related to each other within the cluster, rather than the clusters of genes, is a more favorite subject. That is, it is also important for the algorithm to make graphical presentations not just clusters. K-means, self-organizing maps (SOM), hierarchical clustering, graph-theoretic approach, model-based clustering, and density-based approach (DHC) are the examples of gene-based clustering algorithms.

The goal of the **sample-based approach** is to find the phenotype structure or the sub-structure of the sample. The phenotypes of the samples studied [67] can only be distinguished by small gene subsets whose expression levels are highly correlated with cluster discrimination. These genes are called informative genes. Other genes in the expression matrix have no role in the decomposition of the samples and are considered noise in the database. Traditional clustering algorithms, such as K-means, SOM, and hierarchical clustering, can be applied directly to clustering samples taking all genes as features. The ratio of the promoter genes to the nonrelated genes (noise ratio) is usually 1:10. This also hinders the reliability of the clustering algorithm. These methods are used to identify the informative genes. Selection of the informative genes is examined in two different categories as supervised and unsupervised. The supervised approach is used in cases where phenotype information such as "patient" and "healthy" is added. In this example, the classifier containing only the informative genes is constructed using this information. The supervised approach is often used by biologists to identify informative genes. In the unsupervised approach, no label specifying the phenotype of the samples is placed. The lack of labeling and therefore the fact that the informative genes do not guide clustering makes the unsupervised approach more complicated. There are two problems that need to be addressed in the unsupervised approach: (i) the high number of genes versus the limited number of samples and (ii) the vast majority of collected genes are irrelevant. Two strategies can be mentioned for these problems in the unsupervised approach: unsupervised gene selection and clustering. In unsupervised gene selection, gene selection and sample clustering are treated as two separate processes. First, the gene size is reduced, and then classical clustering algorithms are applied. Since there is no training set, the choice of gene is based solely on statistical models that analyze the variance of gene expression data. Associated clustering dynamically supports the combination of repetitive clustering and gene selection processes by the use of the relationship between genes and samples. After many repetitions, the sample fragments converge to the real sample structure and the selected genes are likely candidates for the informative gene cluster.

When **subspace clustering** is applied to gene expression vectors, it is treated as a "block" consisting of clusters of genes and subclasses of experimental conditions. The expression pattern of the genes in the same block is consistent under the condition in that block. Different greedy heuristic approaches have been adapted to approximate optimal solution.

Subspace clustering was first described by Agrawal et al. in 1998 on general data mining [68]. In subspace clustering, two subspace sets may share the same objects and properties, while some objects may not belong to any subspace set. Subspace clustering methods usually define a model to determine the target block and then search in the gen-sample space. Some examples of subspatial cluster methods proposed for gene expression are biclustering [69], coupled two way clustering (CTWC) [70], and plaid model [71].

According to different clustering criteria, data can be clustered such as the co-expressing gene groups, the samples belonging to the same phenotype or genes from the same biological process. However, even if the same criteria are used in different clustering algorithms, the data can be clustered in different forms. For this reason, it is necessary to select more suitable algorithm for data distribution.

4. Conclusions

Clustering for time-series data is used as an effective method for data analysis of many areas from social media usage and financial data to bioinformatics. There are various methods introduced for time-series data. Which approach is chosen is specific to the application. The application is determined by the needs such as time, speed, reliability, storage, and so on. When determining the approach to clustering, three basic issues need to be decided: data representation, similarity measure, and clustering algorithm.

The data representation involves transforming the multi-dimensional and noisy structure of the time-series data into a less dimensional that best expresses the whole data. The most commonly used method for this purpose is dimension reduction or feature extraction.

It is challenging to measure the similarity of two time series. The chapter has been examined similarity measures in three sections as similarity in shape, similarity in time, and similarity in change.

For the time-series clustering algorithms, it is not wrong to say that the evolution of conventional clustering algorithms. Therefore, the classification of traditional clustering algorithms (developed for static data) has been included. It is classified as partitioning, hierarchical, model-based, grid-based, and density-based. Partition algorithms initially require prototypes. The accuracy of the algorithm depends on the defined prototype and updated method. However, they are successful in finding similar series and clustering time series with equal length. The fact that the number of clusters is not given as the initial parameter is a prominent and well-known feature of hierarchical algorithms. At the same time, works on time series that are not of equal length causes it to be one step ahead of other algorithms. However, hierarchical algorithms are not suitable for large data sets due to the complexity of the calculation and the scalability problem. Model-based algorithms suffer from problems such as initialization of parameters based on user predictions and slow processing time for large databases. Density-based algorithms are not generally preferred over time-series data due to their high working complexity. Each approach has pros and cons compared to each other, and the choice of algorithm for time-series clustering varies completely according to the characteristics of the data and the needs of the application. Therefore, in the last chapter, a study on the clustering of gene expression data, which is a specific field of application, has been mentioned.

In time-series data clustering, there is a need for algorithms that execute fast, accurate, and with less memory on large data sets that can meet today's needs.

Author details

Esma Ergüner Özkoç
Başkent University, Ankara, Turkey

*Address all correspondence to: eeozkoc@baskent.edu.tr

References

[1] Ratanamahatana C. Multimedia retrieval using time series representation and relevance feedback. In: Proceedings of 8th International Conference on Asian Digital Libraries (ICADL2005); 2005. pp. 400-405

[2] Özkoç EE, Oğul H. Content-based search on time-series microarray databases using clustering-based fingerprints. Current Bioinformatics. 2017;**12**(5):398-405. ISSN: 2212-392X

[3] Lin J, Keogh E, Lonardi S, Lankford J, Nystrom D. Visually mining and monitoring massive time series. In: Proceedings of 2004 ACM SIGKDD International Conference on Knowledge Discovery and data Mining–KDD '04; 2004. p. 460

[4] Bornemann L, Bleifuß T, Kalashnikov D, Naumann F, Srivastava D. Data change exploration using time series clustering. Datenbank-Spektrum. 2018;**18**(2):79-87

[5] Rani S, Sikka G. Recent techniques of clustering of time series data: A survey. International Journal of Computers and Applications. 2012;**52**(15):1

[6] Aghabozorgi S, Shirkhorshidi AS, Wah TY. Time-series clustering–A decade review. Information Systems. 2015;**53**:16-38

[7] Lin J, Keogh E, Lonardi S, Chiu B. A symbolic representation of time series, with implications for streaming algorithms. In: Proceedings of the 8th ACM SIGMOD Workshop on Research Issues in Data Mining and Knowledge Discovery; 13 June 2003; ACM; pp. 2-11

[8] Keogh EJ, Pazzani MJ. A simple dimensionality reduction technique for fast similarity search in large time series databases. In: Pacific-Asia Conference on Knowledge Discovery and Data Mining; 18 April 2000; Springer, Berlin, Heidelberg. pp. 122-133

[9] Esling P, Agon C. Time-series data mining. ACM Computing Surveys (CSUR). 2012;**45**(1):12

[10] Keogh E, Lin J, Fu A. Hot sax: Efficiently finding the most unusual time series subsequence. In: Fifth IEEE International Conference on Data Mining (ICDM'05); 27 November 2005; IEEE. pp. 226-233

[11] Ghysels E, Santa-Clara P, Valkanov R. Predicting volatility: Getting the most out of return data sampled at different frequencies. Journal of Econometrics. 2006;**131**(1-2):59-95

[12] Kawagoe GD. Grid Representation of Time Series Data for Similarity Search. In: Data Engineering Workshop; 2006

[13] Agronomischer Zeitreihen CA. Time Series Clustering in the Field of Agronomy. Technische Universitat Darmstadt (Master-Thesis); 2013

[14] Keogh E, Lonardi S, Ratanamahatana C. Towards parameter-free data mining. In: Proceedings of Tenth ACM SIGKDD International Conference on Knowledge Discovery Data Mining; 2004, Vol. 22, No. 25. pp. 206-215

[15] Keogh E, Chakrabarti K, Pazzani M, Mehrotra S. Locally adaptive dimensionality reduction for indexing large time series databases. ACM SIGMOD Record. 2001;**27**(2):151-162

[16] Keogh E, Pazzani M. An enhanced representation of time series which allows fast and accurate classification, clustering and relevance feedback. In: Proceedings of the 4th International

Conference of Knowledge Discovery and Data Mining; 1998. pp. 239-241

[17] Korn F, Jagadish HV, Faloutsos C. Efficientlysupportingadhoc queries in large datasets of time sequences. ACM SIGMOD Record. 1997;**26**: 289-300

[18] Faloutsos C, Ranganathan M, Manolopoulos Y. Fasts ubsequence matching in time-series databases. ACM SIGMOD Record. 1994;**23**(2): 419-429

[19] Portet F, Reiter E, Gatt A, Hunter J, Sripada S, Freer Y, et al. Automatic generation of textual summaries from neonatal intensive care data. Artificial Intelligence. 2009;**173**(7):789-816

[20] Chan K, Fu AW. Efficient time series matching by wavelets. In: Proceedings of 1999 15th International Conference on Data Engineering; 1999, Vol. 15, no. 3. pp. 126-133

[21] Agrawal R, Faloutsos C, Swami A. Efficient similarity search in sequence databases. Foundations of Data Organization and Algorithms. 1993;**46**: 69-84

[22] Kawagoe K, Ueda T. A similarity search method of time series data with combination of Fourier and wavelet transforms. In: Proceedings Ninth International Symposium on Temporal Representation and Reasoning; 2002. pp. 86-92

[23] Chung FL, Fu TC, Luk R. Flexible time series pattern matching based on perceptually important points. In: Jt. Conference on Artificial Intelligence Workshop. 2001. pp. 1-7

[24] Keogh E, Pazzani M, Chakrabarti K, Mehrotra S. A simple dimensionality reduction technique for fast similarity search in large time series databases. Knowledge and Information Systems. 2000;**1805**(1):122-133

[25] Caiand Y, Ng R. Indexing spatio-temporal trajectories with Chebyshev polynomials. In: Procedings of 2004 ACM SIGMOD International; 2004. p. 599

[26] Bingham E. Random projection in dimensionality reduction: Applications to image and text data. In: Proceedings of the Seventh ACM SIGKDD International Conference on Knowledge Discovery and Data Mining; 2001. pp. 245-250

[27] Chen Q, Chen L, Lian X, Liu Y. Indexable PLA for efficient similarity search. In: Proceedings of the 33rd International Conference on Very large Data Bases; 2007. pp. 435-446

[28] Corduas M, Piccolo D. Timeseries clustering and classification by the autoregressive metric. Computational Statistics & Data Analysis. 2008;**52**(4): 1860-1872

[29] Kalpakis K, Gada D, Puttagunta V. Distance measures for effective clustering of ARIMA time-series. In: Proceedings 2001 IEEE International Conference on Data Mining; 2001. pp. 273-280

[30] Kumar N, Lolla N, Keogh E, Lonardi S. Time-series bitmaps: A practical visualization tool for working with large time series databases. In: Proceedings of the 2005 SIAM International Conference on Data Mining; 2005. pp. 531-535

[31] Minnen D, Starner T, Essa M, Isbell C. Discovering characteristic actions from on body sensor data. In: Proceedings of 10th IEEE International Symposium on Wearable Computers; 2006. pp. 11-18

[32] Minnen D, Isbell CL, Essa I, Starner T. Discovering multivariate motifs using

subsequence density estimation and greedy mixture learning. In: Proceedings of the National Conference on Artificial Intelligence; 2007, Vol. 22, No. 1. p. 615

[33] Panuccio A, Bicego M, Murino V. A hidden Markov model-based approach to sequential data clustering. In: Joint IAPR International Workshops on Statistical Techniques in Pattern Recognition (SPR) and Structural and Syntactic Pattern Recognition (SSPR). Berlin, Heidelberg: Springer; 2002, pp. 734-743

[34] Bagnall AAJ, "Ann" Ratanamahatana C, Keogh E, Lonardi S, Janacek G. A bit level representation for time series data mining with shape based similarity. Data Mining and Knowledge Discovery. 2006;**13**(1): 11-40

[35] Ratanamahatana C, Keogh E, Bagnall AJ, Lonardi S. A novel bit level time series representation with implications for similarity search and clustering. In: Proceedings of 9th Pacific-Asian International Conference on Knowledge Discovery and Data Mining (PAKDD'05); 2005. pp. 771-777

[36] Bagnall AJ, Janacek G. Clustering time series with clipped data. Machine Learning. 2005;**58**(2):151-178

[37] Sakoe H, Chiba S. A dynamic programming approach to continuous speech recognition. In: Proceedings of the Seventh International Congress on Acousticsvol; 1971, Vol. 3. pp. 65-69

[38] Sakoe H, Chiba S. Dynamic programming algorithm optimization for spoken word recognition. IEEE Transactions on Acoustics, Speech, and Signal Processing. 1978;**26**(1):43-49

[39] Smyth P. Clustering sequences with hidden Markov models. Advances in Neural Information Processing Systems. 1997;**9**:648-654

[40] Xiong Y, Yeung DY. Mixtures of ARMA models for model-based time series clustering. In: Data Mining, 2002. ICDM 2003; 2002. pp. 717-720

[41] Han J, Kamber M. Data Mining: Concepts and Techniques. San Francisco: Morgan Kaufmann; 2001. pp. 346-389

[42] Liao TW. Clustering of time series data—a survey. Pattern Recognition. 2005;**38**(11):1857-1874

[43] MacQueen J. Some methods for classification and analysis of multivariate observations. In: Proceedings of the Fifth Berkeley Symposium on Mathematical Statistics and Probability; 21 June 1967, Vol. 1, No. 14. pp. 281-297

[44] Tamayo P, Slonim D, Mesirov J, Zhu Q, Kitareewan S, Dmitrovsky E, et al. Interpreting patterns of gene expression with self-organizing maps: Methods and application to hematopoietic differentiation. Proceedings of the National Academy of Sciences. 1999;**96**(6):2907-2912

[45] Ng RT, Han J. Efficient and effective clustering methods for spatial data mining. In: Proceedings of the International Conference on Very Large Data Bases; 1994. pp. 144-144

[46] Kaufman L, Rousseeuw PJ, Corporation E. Finding Groups in Data: An Introduction to Cluster Analysis, Vol. 39. Hoboken, NewJersey: Wiley Online Library; 1990

[47] Guha S, Rastogi R, Shim K. CURE: An efficient clustering algorithm for large databases. ACM SIGMOD Record. 1998;**27**(2):73-84

[48] Zhang T, Ramakrishnan R, Livny M. BIRCH: An efficient data clustering method for very large databases. ACM SIGMOD Record. 1996;**25**(2): 103-114

[49] Karypis G, Han EH, Kumar V. Chameleon: Hierarchical clustering using dynamic modeling. Computer. 1999;**32**(8):68-75

[50] Beal M, Krishnamurthy P. Gene expression time course clustering with countably infinite hidden Markov models. arXiv preprint arXiv:1206.6824; 2012

[51] Ester M, Kriegel HP, Sander J, Xu X. A density-based algorithm for discovering clusters in large spatial data bases with noise. In: Knowledge Discovery and Data Mining. Vol. 96, No. 34; August 1996. pp. 226-231

[52] Ankerst M, Breunig M, Kriegel H. OPTICS: Ordering points to identify the clustering structure. ACM SIGMOD Record. 1999;**28**(2):40-60

[53] Fisher DH. Knowledge acquisition via incremental conceptual clustering. Machine Learning. 1987;**2**(2):139-172

[54] Carpenter GA, Grossberg S. A massively parallel architecture for a self-organizing neural pattern recognition machine. Computer Vision Graphics Image Process. 1987;**37**(1):54-115

[55] Bouveyron C, Côme E, Jacques J. The discriminative functional mixture model for the analysis of bike sharing systems. The Annals of Applied Statistics. 2015;**9**(4):1726-1760

[56] Wang W, Yang J, Muntz R. STING: A statistical information grid approach to spatial data mining. In: Proceedings of the International Conference on Very Large Data Bases; 1997. pp. 186-195

[57] Bouveyron C, Brunet C. Simultaneous model-based clustering and visualization in the fisher discriminative subspace. Statistics and Computing. 2012;**22**:301-324

[58] Akaike H. A new look at the statistical model identification. IEEE Transactions on Automatic Control. 1974;**19**:716-723

[59] Kass RE, Raftery AE. Bayes factors. Journal of the American Statistical Association. 1995;**90**(430):773-795

[60] Biernacki C, Celeux G, Govaert G. Assessing a mixture model for clustering with the integrated completed likelihood. IEEE Transactions on Pattern Analysis and Machine Intelligence. 2000;**22**:719-725

[61] Bouveyron C. funFEM: Clustering in the Discriminative Functional Subspace. R package version. 2015;1

[62] Barrett T, Troup DB, Wilhite SE, Ledoux P, Rudnev D, Evangelista C, et al. NCBI GEO: Archive for high-throughput functional genomic data. Nucleic Acids Research. 2009;**37** (Database):D885-D890

[63] Kuenzel L. Gene clustering methods for time series microarray data. Biochemistry. 2010;**218**

[64] Moller-Levet CS, Cho KH, Yin H, Wolkenhauer O. Clustering of gene expression time-series data. Technical report. Department of Computer Science, University of Rostock, Germany; 2003

[65] Beal M, Krishneamurthy P. Gene expression time course clustering with countably infinite hidden Markov models. arXiv preprint arXiv:1206.6824; 2012

[66] Jiang D, Tang C, Zhang A. Cluster analysis for gene expression data: A survey. IEEE Transactions on Knowledge and Data Engineering. 2004; **16**(11):1370-1386

[67] Golub TR, Slonim DK, Tamayo P, Huard C, Gaasenbeek M, Mesirov JP, et al. Molecular classification of cancer: Class discovery and class prediction by

gene expression monitoring. Science. 1999;**286**(5439):531-537

[68] Agrawal R, Gehrke J, Gunopulos D, Raghavan P. Automatic Subspace Clustering of High Dimensional Data for Data Mining Applications. ACM; 1998; **27**(2):94-105

[69] Cheng Y, Church GM. Biclustering of expression data. In: ISMB; 2000, Vol. 8, No. 2000. pp. 93-103

[70] Getz G, Levine E, Domany E. Coupled two-way clustering analysis of gene microarray data. Proceedings of the National Academy of Sciences. 2000;**97**(22):12079-12084

[71] Lazzeroni L, Owen A. Plaid models for gene expression data. Statistica Sinica. 2002;**1**:61-86

Chapter 7

Weather Nowcasting Using Deep Learning Techniques

Makhamisa Senekane, Mhlambululi Mafu and Molibeli Benedict Taele

Abstract

Weather variations play a significant role in peoples' short-term, medium-term or long-term planning. Therefore, understanding of weather patterns has become very important in decision making. Short-term weather forecasting (nowcasting) involves the prediction of weather over a short period of time; typically few hours. Different techniques have been proposed for short-term weather forecasting. Traditional techniques used for nowcasting are highly parametric, and hence complex. Recently, there has been a shift towards the use of artificial intelligence techniques for weather nowcasting. These include the use of machine learning techniques such as artificial neural networks. In this chapter, we report the use of deep learning techniques for weather nowcasting. Deep learning techniques were tested on meteorological data. Three deep learning techniques, namely multilayer perceptron, Elman recurrent neural networks and Jordan recurrent neural networks, were used in this work. Multilayer perceptron models achieved 91 and 75% accuracies for sunshine forecasting and precipitation forecasting respectively, Elman recurrent neural network models achieved accuracies of 96 and 97% for sunshine and precipitation forecasting respectively, while Jordan recurrent neural network models achieved accuracies of 97 and 97% for sunshine and precipitation nowcasting respectively. The results obtained underline the utility of using deep learning for weather nowcasting.

Keywords: nowcasting, deep learning, artificial neural network, Elman network, Jordan network, precipitation, rainfall

1. Introduction

Weather changes play a significant role in peoples' short-term, medium-term or long-term planning. Therefore, the understanding weather patterns have become very important in decision making. This further raises the need for availability of tools for accurate prediction of weather. This need is even more pronounced if the prediction is intended for a short-term weather forecasting, conventionally known as nowcasting.

To date, different weather nowcasting models have been proposed [1]. These models are mainly based on the different variants of artificial neural networks and fuzzy logic. As will be discussed later in this chapter, these techniques have some limitations which need to be addressed. In this chapter, we report the use multilayer perceptron (MLP) neural networks, Elman neural networks (ENN) and Jordan

 IntechOpen

neural networks for solar irradiance (sunshine) and precipitation (rainfall) nowcasting. The approach taken in this work is in line with the observation given in [1] in the sense that the performances of these models are further compared in order to establish which model performs best in weather nowcasting. The main contribution of the work reported in this chapter is the development of three solar irradiation and rainfall models using MLP, ENN and Jordan neural networks. These three models are examples of deep learning [2–4]. Therefore, the contribution of this work can be summarized as the use of deep learning models for weather nowcasting. Thus, the research question being addressed in this chapter is the design of integrated high-accuracy nowcasting techniques. Furthermore, the objectives of this work include:

- The design of integrated high-accuracy nowcasting techniques using the following deep learning architectures:
  - MLP
  - ENN
  - Jordan recurrent neural networks
- Application of such techniques to the following tasks:
  - sunshine nowcasting
  - precipitation nowcasting

The remainder of this chapter is divided as follows. The next section provides a background information on artificial neural networks and the related work on the use of artificial neural networks in weather nowcasting. This is followed by Section 3, which discusses the method used for the design and Implementation of both solar irradiation and rainfall nowcasting models. Results are provided and discussed in Section 4, while Section 5 concludes this chapter.

2. Preliminaries

2.1 Artificial neural networks (ANNs)

Artificial neural network (ANN) is an example of supervised machine learning [5–7]. It draws inspiration from how the biological neuron in the brain operates. Thus, it mimics natural intelligence in its learning from experience [5]. As a supervised learning algorithm, ANN learns from the examples by constructing an input-output mapping [8]. A typical ANN consists of an input layer, an output layer, and at least one hidden layer. Each layer consists of nodes representing neurons and is connected by weights. Each internal node of artificial neural network consists of two functions, namely transfer function and activation function [6, 7]. The transfer function is a function of inputs (x_i) and weights (w_i), and is given as

$$f(x) = \Sigma w_i x_i + b_i, \tag{1}$$

where b_i is a bias value. On the other hand, an activation function φ is nonlinear and hence responsible for modeling nonlinear relationships. Additionally, this function is differentiable [8]. The output of such an internal node is given as

$$y_i = \varphi(f(x)). \tag{2}$$

Figure 1 shows a schematic diagram of a typical ANN. Each node in the figure represents a neuron, while the arrows represent the weights. The first layer is the input layer, and each node (neuron) of the input layer corresponds to the feature used for prediction. Thus, in **Figure 1**, there would be three features used for prediction. The hidden layer is between the input layer and the output layer. Its nodes take a set of weighted inputs defined by transfer function in Eq. (1), and produces the output given by Eq. (2).

Depending on the number of hidden layers, artificial neural networks can be classified as either shallow neural networks or deep neural networks. In the former class (shallow neural networks), fewer hidden layers are used while on the latter (deep neural networks), several hidden layers are used for better prediction accuracy. Examples of deep neural network architectures include multilayer perceptrons, convolutional neural networks and recurrent neural networks [2, 4, 9]. It is worth noting that both Elman neural networks and Jordan neural networks are examples of recurrent neural networks [4, 10].

2.2 Related work: weather forecasting using ANNs

Different neural networks-based approaches to short-term weather forecasting have been proposed in literature [1, 10, 11]. furthermore, Mellit *et al.* [12] proposed artificial neural network model for predicting global solar radiation. The model

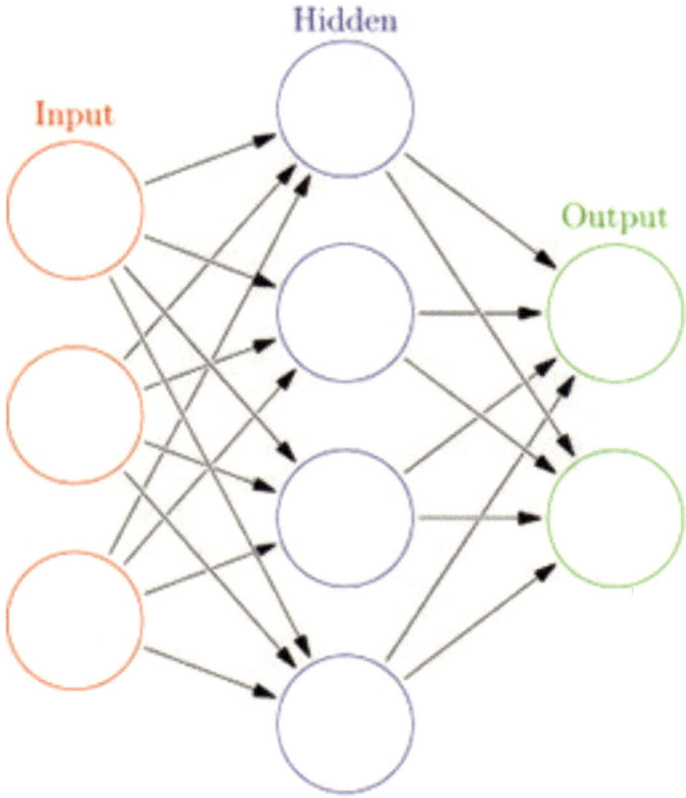

Figure 1.
Schematic diagram of a typical ANN. It consists of an input layer, a hidden layer and an output layer. An input layer consists of three nodes, a hidden layer consists of four nodes, while an output layer consists of two nodes. Since an output layer has two nodes, this ANN is used for two-class (binary) classification. The arrows represent the weights.

proposed uses radial basis function networks, and uses sunshine duration and air temperature as inputs. The model used 300 data points for training, while 65 data points were used for validation and testing. The authors reported that the best performance was obtained with one hidden layer containing 9 neurons (nodes). In Ref. [13], authors reported adaptive neuro-fuzzy inference scheme (ANFIS)-based total solar radiation data forecasting model that takes as inputs daily sunshine duration and mean ambient temperature. The data used in the study spanned a period of 10 years; from 1981 to 1990. It reported validation mean relative error of 1% and correlation coefficient obtained from validation data set was reported to be 98%.

A solar radiation forecasting model based on meteorological data using artificial neural networks is reported in [8]. This algorithm uses meteorological data from Dezful city in Iran. Daily meteorological data from 2002 to 2005 is used to train the model, while 235 days' data from 2006 is used as a testing data. The model takes as inputs length of the day, daily mean air temperature, humidity and sunshine hours. The model achieved absolute testing error of 8.84%. Additionally, Ruffing and Venayagamoorthy [14] proposed a short-to-medium range solar irradiance prediction model using echo state network (ESN). ESN is another variant of a recurrent neural network. The model reported in [14] is capable of predicting solar irradiance 30–270 minutes into the future. Correlation coefficient was used as a performance metric for the model. For 30 minutes ahead predictions, coefficient of correlation was obtained to be 0.87, while for 270 minutes ahead predictions, it decreased to 0.48. Finally, Hosssain *et al.* [15] reported the use of deep learning to forecast weather in Nevada. Their proposed model uses deep neural networks with stacked auto-encoders to predict air temperature using pressure, humidity, wind speed and temperature as inputs. Data used was collected in an hourly interval from November 2013 to December 2014. The model achieved accuracy of 97.94%.

Precipitation forecasting model using artificial neural networks was reported in [16]. This model is capable of estimating 6 hour rainfall over the south coast of Tasmania, Australia. The data used for this model consists of 1000 training examples, 300 validation examples and 300 test examples. The model obtained accuracy of 84%. On the other hand, Shi *et al.* [17] reported a model for precipitation nowcasting which uses convolutional long short term memory (LSTM) network. Unlike other models that were above, this model uses radio detection and ranging (RADAR) data instead of meteorological data. The model obtained a correlation coefficient of 0.908 and mean square error of 1.420.

As will be discussed in the next section, the abovementioned methods are limited compared to the method proposed in this chapter. One of the limitations is that the methods use fewer data instances than the one used in the method discussed in the next section. Another limitation is that the abovementioned techniques use fewer features (less than six features). Finally, unlike the technique discussed in the next section, which integrates different deep learning architectures for both sunshine nowcasting and precipitation nowcasting, the techniques mentioned above either use only one neural network architecture or are designed for one nowcasting task.

3. Methodology for design and implementation of weather nowcasting models

The forecasting models reported in this chapter are tested on hourly weather data from Lesotho for the period ranging from 01/01/2012 to 26/03/2012. This meteorological data consists of 2045 instances; and six features were used to make predictions. As opposed to the approaches discussed in Section 2.2 above, the method discussed in this chapter has two major advantages. The first one is that it uses more data; 2045 instances. Additionally, the method is feature-rich, since it uses six features (more than what other methods reported in Section 2.2 use) for short-term weather forecasting. As a means of feature engineering, all the predictors (features) were plotted against one another, in order to ensure that they are not linearly related, in which case it would be sufficient to use one instead of all those that are related. Therefore, these six features were selected because they proved to be independent predictors. These features are summarized in **Figure 2**. As can be observed from **Figure 2**, all the six features form the nodes of the input layers of all three deep learning architectures (namely, MLP, ENN and Jordan recurrent neural networks). **Figure 3** summarizes the design of the method discussed in this chapter.

The models were developed using R statistical programming language [18–20], and RSNNS package was used to implement artificial neural networks [21]. The models created make use of multilayer perceptron, Elman recurrent neural network and Jordan recurrent neural network. These models were then used for weather Nowcasting to perform two tasks, namely sunshine predictions and precipitation predictions. Additionally, each model was designed with a time lag of 1 hour (thereby allowing 1 hour ahead forecasting). Furthermore, from the collected meteorological data, 80% of the data was used to train the model, 10%

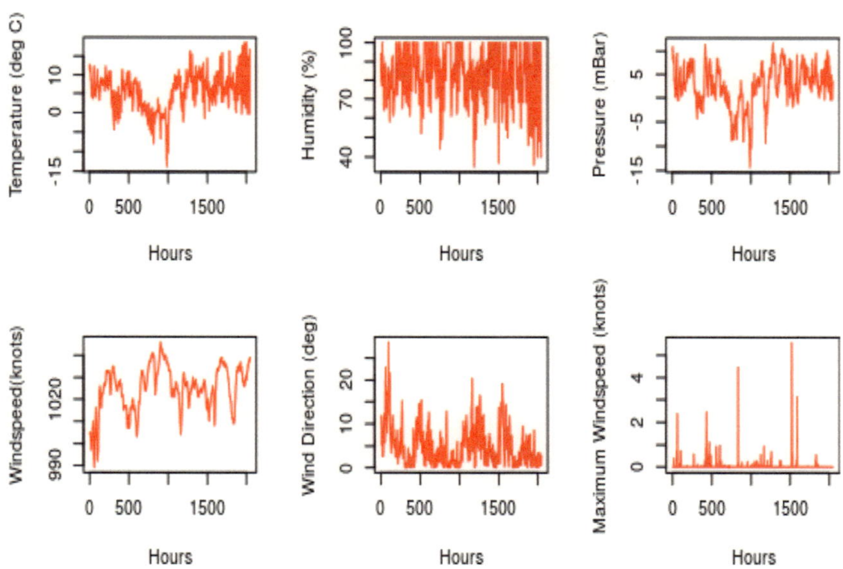

Figure 2.
Summary of features that were used for weather forecasting.

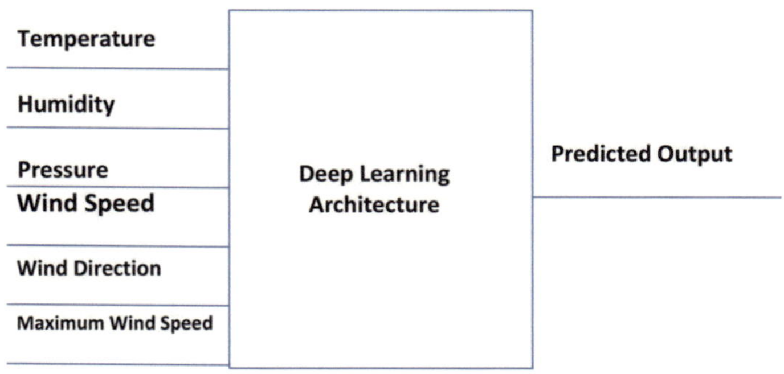

Figure 3.
Summary of the method used for weather nowcasting tasks using deep learning architectures.

for validation, while the remaining 10% was used to test the model for accuracy. In order to enable reproducibility of the result, a seed was set to 2017 using the R command: "set.seed (2017)."

4. Results and discussion

Figures 4–6 show the MLP, Elman RNN and Jordan RNN sunshine forecasting models respectively. The black line is a fit for an ideal model, while a red line is a fit of the proposed model. As it can be observed, Jordan neural network model outperforms the other two models, while the multilayer perceptron model is the poorest of the three in sunshine nowcasting. Additionally, performances of both Elman neural network model and Jordan neural network model are comparable.

Figures 7–9 compare the performances of the three neural network models in precipitation nowcasting. Once again, the black line is a fit for an ideal model, while a red line is a fit of the proposed model. It can be observed that once again, MLP has the lowest performance while Jordan neural network model is the best-performing model. Also, the performances of both the Elman neural network model and Jordan neural network model are comparable.

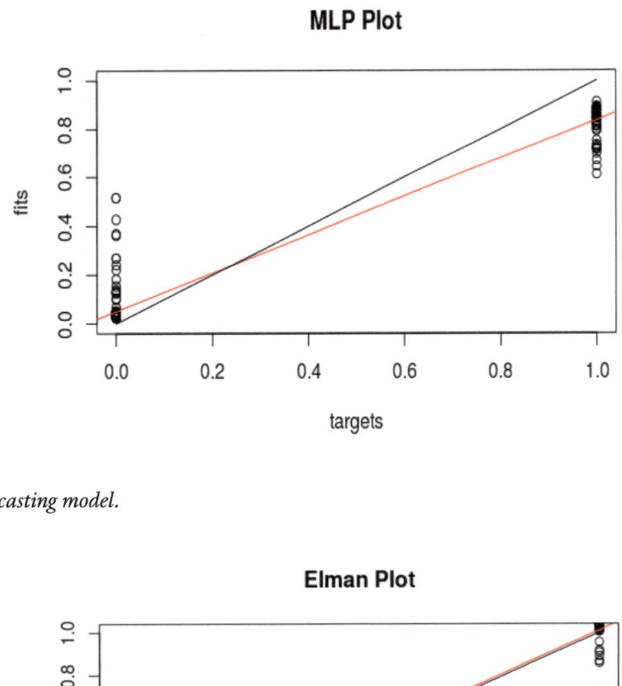

Figure 4.
MLP sunshine forecasting model.

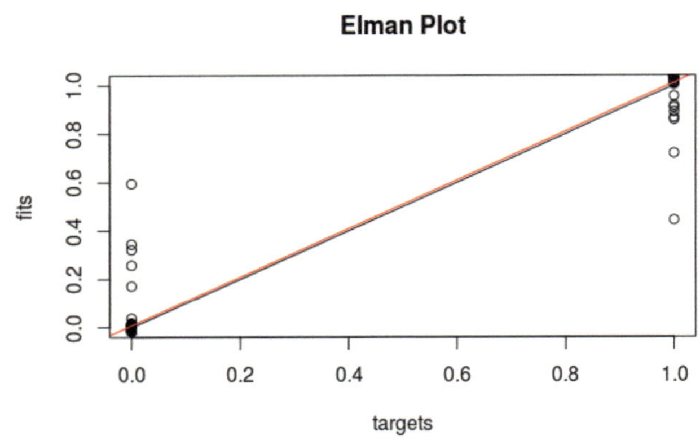

Figure 5.
Elman RNN sunshine forecasting model.

Finally, accuracies of the models were compared, and the results are shown in **Figure 10**. The figure shows that MLP yet once again performing poorly compared to Elman RNN and Jordan RNN. Although these models have high accuracies individually, it is suggested that combining them together (as ensemble of models) might improve the accuracy even further. This possibility warrants further investigation.

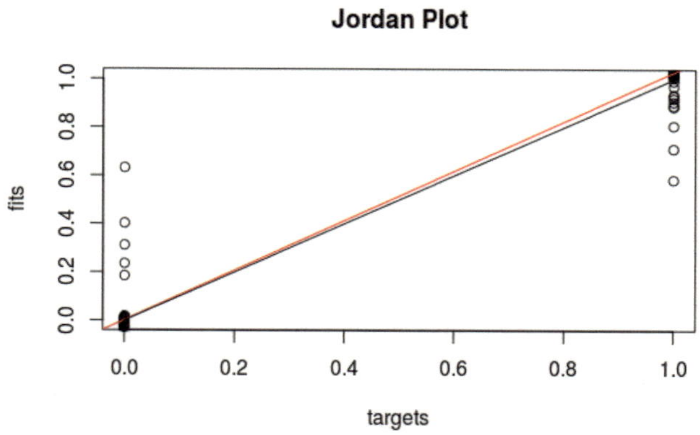

Figure 6.
Jordan RNN sunshine forecasting model.

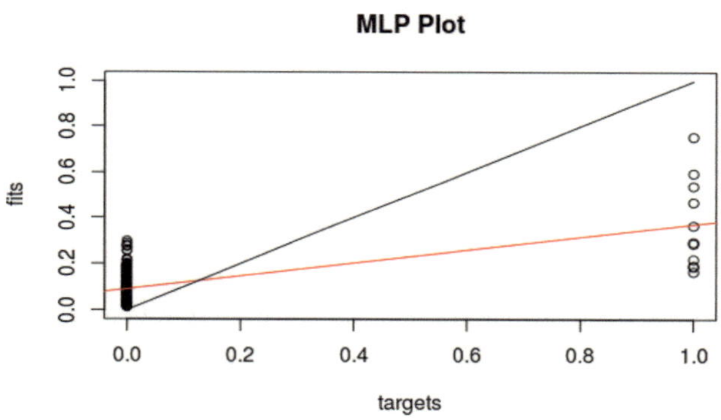

Figure 7.
MLP precipitation forecasting model.

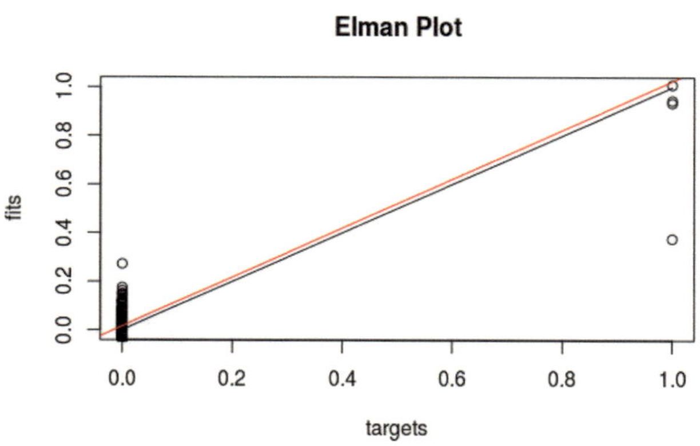

Figure 8.
Elman RNN precipitation forecasting model.

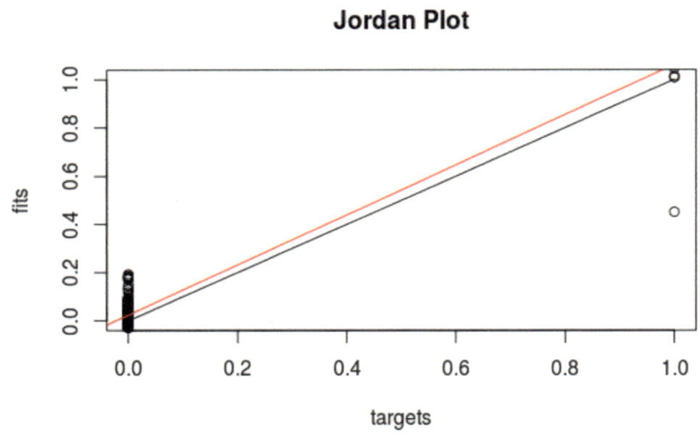

Figure 9.
Jordan RNN precipitation forecasting model.

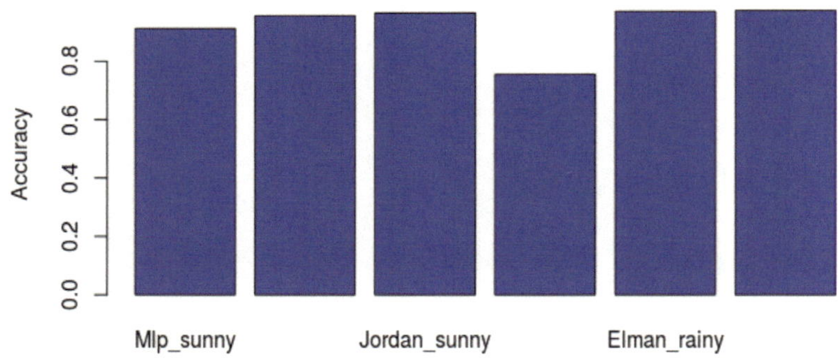

Figure 10.
Accuracies of different neural network models for weather nowcasting.

5. Conclusion

In this chapter, we have reported the application of deep learning for short- term forecasting of Lesotho's weather. The deep learning models used are multilayer perceptron, Elman recurrent neural networks and Jordan neural networks. These models were used to predict sunshine and precipitation. High accuracies of these models in weather forecasting underline their utility. Thus, high-accuracy results obtained from this work, coupled with the integrated nature of the technique reported, provide more advantages over other approaches used for weather nowcasting. Future work will focus on improving the accuracy of weather nowcasting by using an ensemble of the stated deep learning models, instead of using them as individual models.

Acknowledgements

Makhamisa Senekane and Benedict Molibeli Taele acknowledge support from the National University of Lesotho Research and Conferences Committee. Mhlambululi Mafu thanks his colleagues at Botswana International University of Science and Technology for their unwavering support and constructive discussions.

Conflict of interest

The authors declare no conflict of interest.

Author details

Makhamisa Senekane[1]*, Mhlambululi Mafu[2] and Molibeli Benedict Taele[1]

1 Department of Physics and Electronics, National University of Lesotho, Roma, Lesotho

2 Department of Physics and Astronomy, Botswana International University of Science and Technology, Palapye, Botswana

*Address all correspondence to: makhamisa12@gmail.com

References

[1] Yadav AK, Chandel S. Solar radiation prediction using artificial neural network techniques: A review. Renewable and Sustainable Energy Reviews. 2013;**33**:772-781

[2] Goodfellow I, Bengio Y, Courville A. Deep Learning. Massachusetts: MIT Press; 2016, Available from: http://www.deeplearningbook.org

[3] Nielsen M. Neural Networks and Deep Learning. California: Determination Press; 2015

[4] Lewis N. Deep Learning Made Easy with R: A Gentle Introduction For Data Science. California: CreateSpace Independent Publishing Platform; 2016

[5] Wasserman PD. Advanced Methods in Neural Computing. New Jersey: John Wiley & Sons, Inc; 1993

[6] Christopher MB. Pattern Recognition and Machine Learning. New York: Springer-Verlag; 2016

[7] Russell SJ, Norvig P. Artificial Intelligence: A Modern Approach. Malaysia: Pearson Education Limited; 2016

[8] Ghanbarzadeh A, Noghrehabadi A, Assareh E, Behrang M. Solar radiation forecasting based on meteorological data using artificial neural networks. In: Industrial Informatics, 2009. INDIN 2009. 7th IEEE International Conference on, IEEE. 2009. pp. 227-231

[9] Schmidhuber J. Deep learning in neural networks: An overview. Neural Networks. 2015;**61**:85-117

[10] Lewis N. Neural Networks for Time Series Forecasting with R: Intuitive Step by Step for Beginners. California: CreateSpace Independent Publishing Platform; 2017

[11] Yadav AK, Malik H, Chandel S. Selection of most relevant input parameters using weka for artificial neural network based solar radiation prediction models. Renewable and Sustainable Energy Reviews. 2014;**31**:509-519

[12] Mellit A, Menghanem M, Bendekhis M. Artificial neural network model for prediction solar radiation data: Application for sizing stand-alone photovoltaic power system. In: Power Engineering Society General Meeting, 2005. IEEE. IEEE; 2005. pp. 40-44

[13] Mellit A, Arab AH, Khorissi N, Salhi H. An anfis-based forecasting for solar radiation data from sunshine duration and ambient temperature. In: Power Engineering Society General Meeting, 2007. IEEE. IEEE; 2007. pp. 1-6

[14] Ruffing SM, Venayagamoorthy GK. Short to medium range time series prediction of solar irradiance using an echo state network. In: Intelligent System Applications to Power Systems, 2009. ISAP'09. 15th International Conference on, IEEE. 2009. pp. 1-6

[15] Hossain M, Rekabdar B, Louis SJ, Dascalu S. Forecasting the weather of nevada: A deep learning approach. In: Neural Networks (IJCNN), 2015 International Joint Conference on, IEEE. 2015. pp. 1-6

[16] McCullagh J, Bluff K, Ebert E. A neural network model for rainfall estimation. In: Artificial Neural Networks and Expert Systems, 1995. Proceedings., Second New Zealand International Two-Stream Conference on, IEEE. 1995. pp. 389-392

[17] Shi X, Chen Z, Wang H, Yeung D-Y, Wong W-K, Woo W-C. Convolutional LSTM network: A machine learning approach for precipitation nowcasting. In: Advances in Neural Information Processing Systems. New York: Curran Associates; 2015. pp. 802-810

[18] Verzani J. Using R for Introductory Statistics. Florida: Chapman & Hall; 2005

[19] Kohl M. Introduction to Statistical Analysis with R. London: Bookboon; 2015

[20] Stowell S. Using R for Statistics. New York: Apress; 2014

[21] Bergmier C, Benitez J. Neural networks in R using the stuttgart neural network simulator: RSNNS. Journal of Statistical Software. 2012;**46**(7):1-26

Chapter 8

Data Mining and Machine Learning for Software Engineering

Elife Ozturk Kiyak

Abstract

Software engineering is one of the most utilizable research areas for data mining. Developers have attempted to improve software quality by mining and analyzing software data. In any phase of software development life cycle (SDLC), while huge amount of data is produced, some design, security, or software problems may occur. In the early phases of software development, analyzing software data helps to handle these problems and lead to more accurate and timely delivery of software projects. Various data mining and machine learning studies have been conducted to deal with software engineering tasks such as defect prediction, effort estimation, etc. This study shows the open issues and presents related solutions and recommendations in software engineering, applying data mining and machine learning techniques.

Keywords: software engineering tasks, data mining, text mining, classification, clustering

1. Introduction

In recent years, researchers in the software engineering (SE) field have turned their interest to data mining (DM) and machine learning (ML)-based studies since collected SE data can be helpful in obtaining new and significant information. Software engineering presents many subjects for research, and data mining can give further insight to support decision-making related to these subjects.

Figure 1 shows the intersection of three main areas: data mining, software engineering, and statistics/math. A large amount of data is collected from organizations during software development and maintenance activities, such as requirement specifications, design diagrams, source codes, bug reports, program versions, and so on. Data mining enables the discovery of useful knowledge and hidden patterns from SE data. Math provides the elementary functions, and statistics determines probability, relationships, and correlation within collected data. Data science, in the center of the diagram, covers different disciplines such as DM, SE, and statistics.

This study presents a comprehensive literature review of existing research and offers an overview of how to approach SE problems using different mining techniques. Up to now, review studies either introduce SE data descriptions [1], explain tools and techniques mostly used by researchers for SE data analysis [2], discuss the role of software engineers [3], or focus only on a specific problem in SE such as defect prediction [4], design pattern [5], or effort estimation [6]. Some existing review articles having the same target [7] are former, and some of them are not

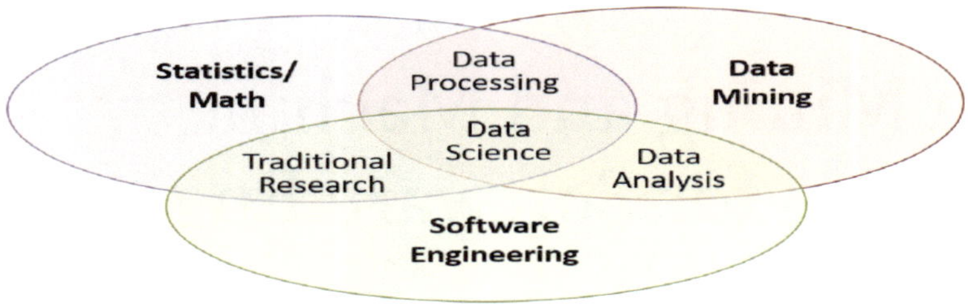

Figure 1.
The intersection of data mining and software engineering with other areas of the field.

comprehensive. In contrast to the previous studies, this article provides a systematic review of several SE tasks, gives a comprehensive list of available studies in the field, clearly states the advantages of mining SE data, and answers "how" and "why" questions in the research area.

The novelties and main contributions of this review paper are fivefold.

- First, it provides a general overview of several SE tasks that have been the focus of studies using DM and ML, namely, defect prediction, effort estimation, vulnerability analysis, refactoring, and design pattern mining.

- Second, it comprehensively discusses existing data mining solutions in software engineering according to various aspects, including methods (clustering, classification, association rule mining, etc.), algorithms (k-nearest neighbor (KNN), neural network (NN), etc.), and performance metrics (accuracy, mean absolute error, etc.).

- Third, it points to several significant research questions that are unanswered in the recent literature as a whole or the answers to which have changed with the technological developments in the field.

- Fourth, some statistics related to the studies between the years of 2010 and 2019 are given from different perspectives: according to their subjects and according to their methods.

- Five, it focuses on different machine learning types: supervised and unsupervised learning, especially on ensemble learning and deep learning.

This paper addresses the following research questions:

RQ1. What kinds of SE problems can ML and DM techniques help to solve?

RQ2. What are the advantages of using DM techniques in SE?

RQ3. Which DM methods and algorithms are commonly used to handle SE tasks?

RQ4. Which performance metrics are generally used to evaluate DM models constructed in SE studies?

RQ5. Which types of machine learning techniques (e.g., ensemble learning, deep learning) are generally preferred for SE problems?

RQ6. Which SE datasets are popular in DM studies?

The remainder of this paper is organized as follows. Section 2 explains the knowledge discovery process that aims to extract interesting, potentially useful, and nontrivial information from software engineering data. Section 3 provides an overview of current work on data mining for software engineering grouped under five tasks: defect prediction, effort estimation, vulnerability analysis, refactoring, and design pattern mining. In addition, some machine learning studies are divided into subgroups, including ensemble learning- and deep learning-based studies. Section 4 gives statistical information about the number of highly validated research conducted in the last decade. Related works considered as fundamental by journals with a highly positive reputation are listed, and the specific methods they used and their categories and purposes are clearly expressed. In addition, widely used datasets related to SE are given. Finally, Section 5 offers concluding remarks and suggests future scientific and practical efforts that might improve the efficiency of SE actions.

2. Knowledge discovery from software engineering data

This section basically explains the consecutive critical steps that should be followed to discover beneficial knowledge from software engineering data. It outlines the order of necessary operations in this process and explains how related data flows among them.

Software development life cycle (SDLC) describes a process to improve the quality of a product in project management. The main phases of SDCL are planning, requirement analysis, designing, coding, testing, and maintenance of a project. In every phase of software development, some software problems (e.g., software bugs, security, or design problems) may occur. Correcting these problems in the early phases leads to more accurate and timely delivery of the project. Therefore, software engineers broadly apply data mining techniques for different SE tasks to solve SE problems and to enhance programming efficiency and quality.

Figure 2 presents the data mining and knowledge discovery process of SE tasks including data collection, data preprocessing, data mining, and evaluation. In the data collection phase, data are obtained from software projects such as bug reports, historical data, version control data, and mailing lists that include various information about the project's versions, status, or improvement. In the data preprocessing phase, the data are preprocessed after collection by using different methods such as feature selection (dimensionality reduction), feature extraction, missing data elimination, class imbalance analysis, normalization, discretization, and so on. In the next phase, DM techniques such as classification, clustering, and association rule mining are applied to discover useful patterns and relationships in software

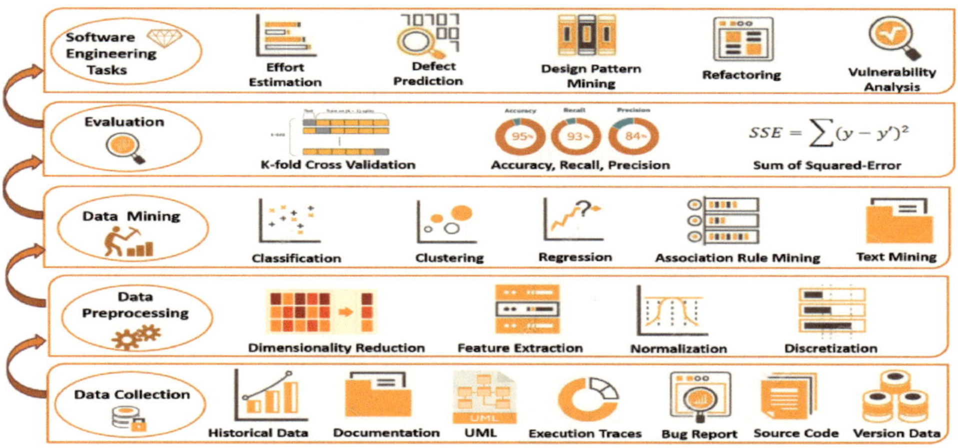

Figure 2.
KDD process for software engineering.

engineering data and therefore to solve a software engineering problem such as defected or vulnerable systems, reused patterns, or parts of code changes. Mining and obtaining valuable knowledge from such data prevents errors and allows software engineers to deliver the project on time. Finally, in the evaluation phase, validation techniques are used to assess the data mining results such as k-fold cross validation for classification. The commonly used evaluation measures are accuracy, precision, recall, F-score, area under the curve (AUC) for classification, and sum of squared errors (SSE) for clustering.

3. Data mining in software engineering

In this review, we examine data mining studies in various SE tasks and evaluate commonly used algorithms and datasets.

3.1 Data mining in defect prediction

A defect means an error, failure, flaw, or bug that causes incorrect or unexpected results in a system [8]. A software system is expected to be without any defects since software quality represents a capacity of the defect-free percentage of the product [9]. However, software projects often do not have enough time or people working on them to extract errors before a product is released. In such a situation, defect prediction methods can help to detect and remove defects in the initial stages of the SDLC and to improve the quality of the software product. In other words, the goal of defect prediction is to produce robust and effective software systems. Hence, software defect prediction (SDP) is an important topic for software engineering because early prediction of software defects could help to reduce development costs and produce more stable software systems.

Various studies have been conducted on defect prediction using different metrics such as code complexity, history-based metrics, object-oriented metrics, and process metrics to construct prediction models [10, 11]. These models can be considered on a cross-project or within-project basis. In within-project defect prediction (WPDP), a model is constructed and applied on the same project [12]. For within-project strategy, a large amount of historical defect data is needed. Hence, in new projects that do not have enough data to train, cross-project strategy may be preferred [13]. Cross-project defect prediction (CPDP) is a method that involves applying a prediction model from one project to another, meaning that models are prepared by utilizing historical data from other projects [14, 15]. Studies in the field of CPDP have increased in recent years [10, 16]. However, there are some deficiencies in comparisons of prior studies since they cannot be replicated because of the difference in utilizing evaluation metrics or preparation way of training data. Therefore, Herbold et al. [16] tried to replicate different CPDP methods previously proposed and find which approach performed best in terms of metrics such as F-score, area under the curve (AUC), and Matthews correlation coefficient (MCC). Results showed that 7- or 8-year approaches may perform better. Another study [17] replicated prior work to demonstrate whether the determination of classification techniques is important. Both noisy and cleaned datasets were used, and the same results were obtained from the two datasets. However, new dataset gave better results for some classification algorithms. For this reason, authors claimed that the selection of classification techniques affects the performance of the model.

Numerous defect prediction studies have been conducted using DM techniques. In the following subsections, we will explain these studies in terms of whether they apply ensemble learning or not. Some defect prediction studies in SE are compared

Ref.	Year	Task	Objective	Algorithms	Ensemble learning	Dataset	Evaluation metrics and results
[18]	2011	Classification	Comparative study of various ensemble methods to find the most effective one	NB	Bagging, boosting, RT, **RF**, RS, AdaBoost, Stacking, and **Voting**	NASA datasets: CM1 JM1 KC1 KC2 KC3 KC4 MC1 MC2 MW1 PC1 PC2 PC3 PC4 PC5	10-fold CV, ACC, and AUC *Vote* 88.48% *random forest* 87.90%
[19]	2013	Classification	Comparative study of class imbalance learning methods and proposed dynamic version of AdaBoost.NC	NB, RUS, RUS-bal, THM, SMB, BNC	RF, SMB, BNC, **AdaBoost.NC**	NASA and PROMISE repository: MC2, KC2, JM1, KC1, PC4, PC3, CM1, KC3, MW1, PC1	10-fold CV Balance, G-mean and AUC, PD, PF
[20]	2014	Classification	Comparative study to deal with imbalanced data	Base Classifiers: C4.5, NB Sampling: ROS, RUS, SMOTE	AdaBoost, Bagging, boosting, RF	NASA datasets: CM1, JM1, KC1, KC2, KC3, MC1, MC2, MW1, PC1, PC2, PC3, PC4, PC5	5 × 5 CV, MCC, ROC, results change according to characteristics of datasets
[17]	2015	Clustering/ classification	To show that the selection of classification technique has an impact on the performance of software defect prediction models	Statistical: NB, **Simple Logistic** Clustering: KM, EM Rule based: Ripper, Ridor NNs: RBF Nearest neighbor: KNN DTs: J48, **LMT**	Bagging, AdaBoost, rotation forest, random subspace	NASA: CM1, JM1, KC1, KC3, KC4, MW1, PC1, PC2, PC3, PC4 PROMISE: Ant 1.7, Camel 1.6, Ivy 1.4, Jedit 4, Log4j 1, Lucene 2.4, Poi 3, Tomcat 6, Xalan 2.6, Xerces 1.3	10 × 10-fold CV AUC > 0.5 Scott-Knott test α = 0.05, simple logistic, LMT, and RF + base learner outperforms KNN and RBF
[21]	2015	Classification	Average probability ensemble (APE) learning module is proposed by combining feature selection and ensemble learning	**APE** system combines seven classifiers: SGD, weighted SVMs (W-SVMs), LR, MNB and Bernoulli naive Bayes (BNB)	RF, GB	NASA: CM1, JM1, KC1, KC3, KC4, MW1, PC1, PC2, PC3, PC4 PROMISE (RQ2): Ant 1.7, Camel 1.6, Ivy 1.4, Jedit 4, Log4j 1, Lucene 2.4, Poi 3, Tomcat 6, Xalan 2.6, Xerces 1.3	10 × 10-fold CV, AUC > 0.5 Scott-Knott test α = 0.05, simple logistic, LMT, and RF + base learner outperforms KNN and RBF
[22, 23]	2016	Classification	Comparative study of 18 ML techniques using OO metrics on six releases of Android operating system	LR, **NB**, BN, **MLP**, RBF SVM, VP, CART, J48, ADT, Nnge, DTNB	Bagging, random forest, Logistic model trees, **Logit Boost**, Ada Boost	6 releases of Android app: Android 2.3.2, Android 2.3.7, Android 4.0.4, Android 4.1.2, Android 4.2.2, Android 4.3.1	10-fold, inter-release validation AUC for NB, LB, MLP is >0.7
[24]	2016	Classification	Caret has been applied whether parameter settings can have a	NB, KNN, LR, partial least squares, NN, LDA, rule based, DT, SVM	Bagging, boosting	Cleaned NASA JM1, PC5 Proprietary from Prop-1 to Prop-5	Out-of-sample bootstrap validation technique, AUC

Ref.	Year	Task	Objective	Algorithms	Ensemble learning	Dataset	Evaluation metrics and results
			large impact on the performance of defect prediction models			Apache Camel 1.2, Xalan 2.5–2.6 Eclipse Platform 2.0–2.1–3.0, Debug 3.4, SWT 3.4, JDT, Mylyn, PDE	Caret AUC performance up to 40 percentage points
[25]	2017	Regression	Aim is to validate the source code metrics and identify a suitable set of source code metrics	5 training algorithms: GD, GDM, GDX, NM, LM	Heterogeneous linear and nonlinear ensemble methods	56 open-source Java projects from PROMISE Repository	10-fold CV, t-test, ULR analysis Neural network with Levenberg Marquardt (LM) is the best
[16]	2017	Classification	Replicate 24 CDPD approaches, and compare on 5 different datasets	DT, LR, NB, SVM	LE, RF, BAG-DT, BAG-NB, BOOST-DT, BOOST-NB	5 available datasets: JURECZKO, NASA MDP, AEEEM, NETGENE, RELINK	Recall, PR, ACC, G-measure, F-score, MCC, AUC
[26]	2017	Classification	Just-in-time defect prediction (TLEL)	NB, SVM, DT, LDA, NN	Bagging, stacking	Bugzilla, Columba, JDT, Platform, Mozilla, and PostgreSQL	10-fold CV, F-score
[13]	2017	Classification	Adaptive Selection of Classifiers in bug prediction (ASCI) method is proposed.	Base classifiers: LOG (binary logistic regression), NB, RBF, MLP, DT	Voting	Ginger Bread (2.3.2 and 2.3.7), Ice Cream Sandwich (4.0.2 and 4.0.4), and JellyBean (4.1.2, 4.2.2 and 4.3.1)	10-fold, inter-release validation AUC for NB, LB, MLP is >0.7
[27]	2018	Classification	MULTI method for JIT-SDP (just in time software defect prediction)	EALR, SL, RBFNet Unsupervised: LT, AGE	Bagging, AdaBoost, Rotation Forest, RS	Bugzilla, Columba, Eclipse JDT, Eclipse Platform, Mozilla, PostgreSQ	CV, timewise-CV, ACC, and P_{OPT} MULTI performs significantly better than all the baselines
[28]	2007	Classification	To found pre- and post-release defects for every package and file	LR	—	Eclipse 2.0, 2.1, 3.0	PR, recall, ACC
[8]	2014	Clustering	Cluster ensemble with PSO for clustering the software modules (fault-prone or not fault-prone)	PSO clustering algorithm	KM-E, KM-M, PSO-E, PSO-M and EM	Nasa MDP, PROMISE	

Ref.	Year	Task	Objective	Algorithms	Ensemble learning	Dataset	Evaluation metrics and results
[29]	2015	Classification	Defect identification by applying DM algorithms	NB, J48, MLP	—	PROMISE, NASA MDP dataset: CM1, JM1, KC1, KC3, MC1, MC2, MW1, PC1, PC2, PC3	10-fold CV, ACC, PR, FMLP is the best
[30]	2015	Classification	To show the attributes that predict the defective state of software modules	NB, NN, association rules, DT	Weighted voting rule of the four algorithms	NASA datasets: CM1, JM1, KC1, KC2, PC1	PR, recall, ACC, F-score NB > NN > DT
[31]	2016	Classification	Authors proposed a model that finds fault-proneness	NB, LR, LivSVM, MLP, SGD, SMO, VP, LR Logit Boost, Decision Stamp, RT, REP Tree	RF	Camel1.6, Tomcat 6.0, Ant 1.7, jEdit4.3, Ivy 2.0, arc, e-learning, berek, forrest 0.8, zuzel, Intercafe, and Nieruchomosci	10-fold CV, AUC AUC = 0.661
[32]	2016	Classification	GA to select suitable source code metrics	LR, ELM, SVML, SVMR, SVMP	—	30 open-source software projects from PROMISE repository from DS1 to DS30	5-fold CV, F-score, ACC, pairwise t-test
[33]	2016	—	Weighted least-squares twin support vector machine (WLSTSVM) to find misclassification cost of DP	SVM, NB, RF, LR, KNN, BN, cost-sensitive neural network	—	PROMISE repository: CM1, KC1, PC1, PC3, PC4, MC2, KC2, KC3	10-fold CV, PR, recall, F-score, G-mean Wilcoxon signed rank test
[34]	2016	—	A multi-objective naive Bayes learning techniques MONB, MOBNN	NB, LR, DT, MODT, MOLR, MONB	—	Jureczko datasets obtained from PROMISE repository	AUC, Wilcoxon rank test CP MO NB (0.72) produces the highest value
[35]	2016	Classification	A software defect prediction model to find faulty components of a software	Hybrid filter approaches FISHER, MR, ANNIGMA.	—	KC1, KC2, JM1, PC1, PC2, PC3, and PC4 datasets	ACC, ent filters, ACC 90%
[36]	2017	Classification	Propose an hybrid method called TSC-RUS + S	A random undersampling based on two-step cluster (TSC)	Stacking: DT, LR, kNN, NB	NASA MDP: i.e., CM1, KC1, KC3, MC2, MW1, PC1, PC2, PC3, PC4	10-fold CV, AUC, (TSC-RUS + S) is the best
[37]	2017	Classification	Analyze five popular ML algorithms for software defect prediction	ANN, PSO, DT, NB, LC	—	Nasa and PROMISE datasets: CM1, JM1, KC1, KC2, PC1, KC1-LC	10-fold CV ANN < DT

Ref.	Year	Task	Objective	Algorithms	Ensemble learning	Dataset	Evaluation metrics and results
[38]	2018	Classification	Three well-known ML techniques are compared.	NB, DT, ANN	—	Three different datasets DS1, DS2, DS3	ACC, PR, recall, F, ROC ACC 97% DT > ANN > NB
[10]	2018	Classification	ML algorithms are compared with CODEP	LR, BN, RBF, MLP, alternating decision tree (ADTree), and DT	Max, CODEP, Bagging J48, Bagging NB, Boosting J48, Boosting NB, RF	PROMISE: Ant, Camel, ivy, Jedit, Log4j, Lucene, Poi, Prop, Tomcat, Xalan	F-score, PR, AUC ROC Max performs better than CODEP

Table 1.
Data mining and machine learning studies on the subject "defect prediction."

in **Table 1**. The objective of the studies, the year they were conducted, algorithms, ensemble learning techniques and datasets in the studies, and the type of data mining tasks are shown in this table. The bold entries in **Table 1** have better performance than other algorithms in that study.

3.1.1 Defect prediction using ensemble learning techniques

Ensemble learning combines several base learning models to obtain better performance than individual models. These base learners can be acquired with:

i. Different learning algorithms

ii. Different parameters of the same algorithm

iii. Different training sets

The commonly used ensemble techniques bagging, boosting, and stacking are shown in **Figure 3** and briefly explained in this part. Bagging (which stands for bootstrap aggregating) is a kind of parallel ensemble. In this method, each model is built independently, and multiple training datasets are generated from the original dataset through random selection of different feature subsets; thus, it aims to decrease variance. It combines the outputs of each ensemble member by a voting mechanism. Boosting can be described as sequential ensemble. First, the same weights are assigned to data instances; after training, the weight of wrong predictions is increased, and this process is repeated as the ensemble size. Finally, it uses a weighted voting scheme, and in this way, it aims to decrease bias. Stacking is a technique that uses predictions from multiple models via a meta-classifier.

Some software defect prediction studies have compared ensemble techniques to determine the best performing one [10, 18, 21, 39, 40]. In a study conducted by Wang et al. [18], different ensemble techniques such as bagging, boosting, random tree, random forest, random subspace, stacking, and voting were compared to each other and a single classifier (NB). According to the results, voting and random forest clearly exhibited better performance than others. In a different study [39],

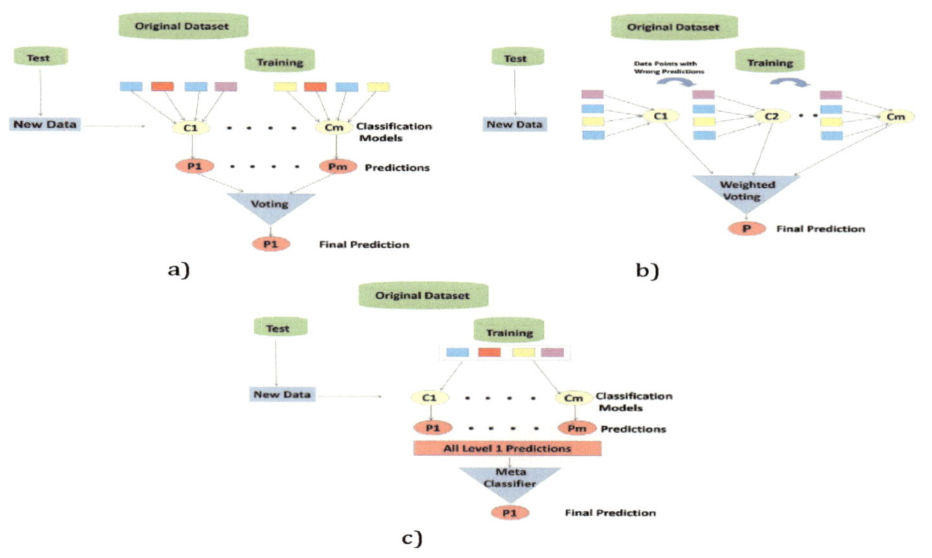

Figure 3.
Common ensemble learning methods: (a) Bagging, (b) boosting, (c) stacking.

ensemble methods were compared with more than one base learner (NB, BN, SMO, PART, J48, RF, random tree, IB1, VFI, DT, NB tree). For boosted SMO, bagging J48, and boosting and bagging RT, performance of base classifiers was lower than that of ensemble learner classifiers.

In study [21], a new method was proposed of mixing feature selection and ensemble learning for defect classification. Results showed that random forests and the proposed algorithm are not affected by poor features, and the proposed algorithm outperforms existing single and ensemble classifiers in terms of classification performance. Another comparative study [10] used seven composite algorithms (Ave, Max, Bagging C4.5, bagging naive Bayes (NB), Boosting J48, Boosting naive Bayes, and RF) and one composite state-of-the art study for cross-project defect prediction. The Max algorithm yielded the best results regarding F-score in terms of classification performance.

Bowes et al. [40] compared RF, NB, Rpart, and SVM algorithms to determine whether these classifiers obtained the same results. The results demonstrated that a unique subset of defects can be discovered by specific classifiers. However, whereas some classifiers are steady in the predictions they make, other classifiers change in their predictions. As a result, ensembles with decision-making without majority voting can perform best.

One of the main problems of SDP is the imbalance between the defect and non-defect classes of the dataset. Generally, the number of defected instances is greater than the number of non-defected instances in the collected data. This situation causes the machine learning algorithms to perform poorly. Wang and Yao [19] compared five class-imbalanced learning methods (RUS, RUS-bal, THM, BNC, SMB) and NB and RF algorithms and proposed the dynamic version of AdaBoost. NC. They utilized balance, G-mean, and AUC measures for comparison. Results showed that AdaBoost.NC and naive Bayes are better than the other seven algorithms in terms of evaluation measures. Dynamic AdaBoost.NC showed better defect detection rate and overall performance than the original AdaBoost.NC. To handle the class imbalance problem, studies [20] have compared different methods (sampling, cost sensitive, hybrid, and ensemble) by taking into account evaluation metrics such as MCC and receiver operating characteristic (ROC).

As shown in **Table 1**, the most common datasets used in the defect prediction studies [17–19, 39] are the NASA MDP dataset and PROMISE repository datasets. In addition, some studies utilized open-source projects such as Bugzilla Columba and Eclipse JDT [26, 27], and other studies used Android application data [22, 23].

3.1.2 Defect prediction studies without ensemble learning

Although use of ensemble learning techniques has dramatically increased recently, studies that do not use ensemble learning are still conducted and successful. For example, in study [32], prediction models were created using source code metrics as in ensemble studies but by using different feature selection techniques such as genetic algorithm (GA).

To overcome the class imbalance problem, Tomar and Agarwal [33] proposed a prediction system that assigns lower cost to non-defective data samples and higher cost to defective samples to balance data distribution. In the absence of enough data within a project, required data can be obtained from cross projects; however, in this case, this situation may cause class imbalance. To solve this problem, Ryu and Baik [34] proposed multi-objective naïve Bayes learning for cross-project environments. To obtain significant software metrics on cloud computing environments, Ali et al. used a combination of filter and wrapper approaches [35]. They compared different machine learning algorithms such as NB, DT, and MLP [29, 37, 38, 41].

3.2 Data mining in effort estimation

Software effort estimation (SEE) is critical for a company because hiring more employees than required will cause loss of revenue, while hiring fewer employees than necessary will result in delays in software project delivery. The estimation analysis helps to predict the amount of effort (in person hours) needed to develop a software product. Basic steps of software estimation can be itemized as follows:

- Determine project objectives and requirements.
- Design the activities.
- Estimate product size and complexity.
- Compare and repeat estimates.

SEE contains requirements and testing besides predicting effort estimation [42]. Many research and review studies have been conducted in the field of SEE. Recently, a survey [43] analyzed effort estimation studies that concentrated on ML techniques and compared them with studies focused on non-ML techniques. According to the survey, case-based reasoning (CBR) and artificial neural network (ANN) were the most widely used techniques. In 2014, Dave and Dutta [44] examined existing studies that focus only on neural network.

The current effort estimation studies using DM and ML techniques are available in **Table 2**. This table summarizes the prominent studies in terms of aspects such as year, data mining task, aim, datasets, and metrics. **Table 2** indicates that neural network is the most widely used technique for the effort estimation task.

Several studies have compared ensemble learning methods with single learning algorithms [45, 46, 48, 49, 51, 60] and examined them on cross-company (CC) and within-company (WC) datasets [50]. The authors observed that ensemble methods obtained by a proper combination of estimation methods achieved better results than single methods. Various ML techniques such as neural network, support vector machine (SVM), and k-nearest neighbor are commonly used as base classifiers for ensemble methods such as bagging and boosting in software effort estimation. Moreover, their results indicate that CC data can increase performance over WC data for estimation techniques [50].

In addition to the abovementioned studies, researchers have conducted studies without using ensemble techniques. The general approach is to investigate which DM technique has the best effect on performance in software effort estimation. For instance, Subitsha and Rajan [54] compared five different algorithms—MLP, RBFNN, SVM, ELM, and PSO-SVM—and Nassif et al. [57] investigated four neural network algorithms—MLP, RBFNN, GRNN, and CCNN. Although neural networks are widely used in this field, missing values and outliers frequently encountered in the training set adversely affect neural network results and cause inaccurate estimations. To overcome this problem, Khatibi et al. [53] split software projects into several groups based on their similarities. In their studies, the C-means clustering algorithm was used to determine the most similar projects and to decrease the impact of unrelated projects, and then analogy-based estimation (ABE) and NN were applied. Another clustering study by Azzeh and Nassif [59] combined SVM and bisecting k-medoids clustering algorithms; an estimation model was then built using RBFNN. The proposed method was trained on historical use case points (UCP).

Ref.	Year	Task	Objective	Algorithms	Ensemble learning	Dataset	Evaluation metrics and results
[45]	2008	Regression	Ensemble of neural networks with associative memory (ENNA)	NN, MLP, KNN	Bagging	NASA, NASA 93, USC, SDR, Desharnais	MMRE, MdMRE and PRED(L) For ENNA PRED(25) = 36.4 For neural network PRED(25) = 8
[46]	2009	Regression	Authors proposed the ensemble of neural networks with associative memory (ENNA)	NN, MLP, KNN	Bagging	NASA, NASA 93, USC, SDR, Desharnais	Random subsampling, t-test MMRE, MdMRE, and PRED(L) ENNA is the best
[47]	2010	Regression	To show the effectiveness of SVR for SEE	SVR, RBF	—	Tukutuku	LOOCV, MMRE, Pred(25), MEMRE, MdEMRE SVR outperforms others
[48]	2011	Regression	To evaluate whether readily available ensemble methods enhance SEE	MLP, RBF, RT	Bagging	5 datasets from PROMISE: cocomo81, nasa93, nasa, sdr, and Desharnais 8 datasets from ISBSG repository	MMRE, MdMRE, PRED(25) RTs and Bagging with MLPs perform similarly
[49]	2012	Regression	To show the measures behave in SEE and to create good ensembles	MLP, RBF, REPTree,	Bagging	cocomo81, nasa93, nasa, cocomo2, desharnais, ISBSG repository	MMRE, PRED(25), LSD, MdMRE, MAE, MdAE Pareto ensemble for all measures, except LSD.
[50]	2012	Regression	To use cross-company models to create diverse ensembles able to dynamically adapt to changes	WC RTs, CC-DWM	WC-DWM	3 datasets from ISBSG repository (ISBSG2000, ISBSG2001, ISBSG) 2 datasets from PROMISE (CocNasaCoc81 and CocNasaCoc81Nasa93)	MAE, Friedman test Only DCL could improve upon RT CC data potentially beneficial for improving SEE
[51]	2012	Regression	To generate estimates from ensembles of multiple prediction methods	CART, NN, LR, PCR, PLSR, SWR, ABE0-1NN, ABE0-5NN	Combining top M solo methods	PROMISE	MAR, MMRE, MdMRE, MMER, MBRE, MIBRE. Combinations perform better than 83%
[52]	2012	Classification/ regression	DM techniques to estimate software effort.	M5, CART, LR, MARS, MLPNN, RBFNN, SVM	—	Coc81, CSC, Desharnais, Cocnasa, Maxwell, USP05	MdMRE, Pred(25), Friedman test Log + OLS > LMS, BC + OLS, MARS, LS-SVM

Ref.	Year	Task	Objective	Algorithms	Ensemble learning	Dataset	Evaluation metrics and results
[53]	2013	Clustering/ classification	Estimation of software development effort	NN, ABE, C-means	—	Maxwell	3-fold CV and LOOCV, RE, MRE, MMRE, PRED
[54]	2014	Regression	ANNs are examined using COCOMO model	MLP, RBFNN, SVM, PSO-SVM Extreme learning Machines	—	COCOMO II Data	MMRE, PRED PSO-SVM is the best
[55]	2014	—	A hybrid model based on GA And ACO for optimization	GA, ACO	—	NASA datasets	MMRE, the proposed method is the best
[56]	2015	Regression	To display the effect of data preprocessing techniques on ML methods in SEE	CBR, ANN, CART Preprocessing rech: MDT, LD, MI, FS, CS, FSS, BSS	—	ISBSG, Desharnais, Kitchenham, USPFT	CV, MBRE, PRED (0.25), MdBRE
[57]	2016	Regression	Four neural network models are compared with each other.	MLP, RBFNN, GRNN, CCNN	—	ISBSG repository	10-fold CV, MAR The CCNN outperforms the other three models
[58]	2016	Regression	To propose a model based on Bayesian network	GA and PSO	—	COCOMO NASA Dataset	DIR, DRM The proposed model is best
[59]	2016	Classification/ regression	A hybrid model using SVM and RBNN compared against previous models	SVM, RBNN	—	Dataset1 = 45 industrial projects Dataset2 = 65 educational projects	LOOCV, MAE, MBRE, MIBRE, SA The proposed approach is the best
[60]	2017	Classification	To estimate software effort by using ML techniques	SVM, KNN	Boosting: kNN and SVM	Desharnais, Maxwell	LOOCV, k-fold CV ACC = 91.35% for Desharnais ACC = 85.48% for Maxwell

Table 2.
Data mining and machine learning studies on the subject "effort estimation."

Zare et al. [58] and Maleki et al. [55] utilized optimization methods for accurate cost estimation. In the former study, a model was proposed based on Bayesian network with genetic algorithm and particle swarm optimization (PSO). The latter study used GA to optimize the effective factors' weight, and then trained by ant colony optimization (ACO). Besides conventional effort estimation studies, researchers have utilized machine learning techniques for web applications. Since web-based software projects are different from traditional projects, the effort estimation process for these studies is more complex.

It is observed that PRED(25) and MMRE are the most popular evaluation metrics in effort estimation. MMRE stands for the mean magnitude relative error, and PRED(25) measures prediction accuracy and provides a percentage of predictions within 25% of actual values.

3.3 Data mining in vulnerability analysis

Vulnerability analysis is becoming the focal point of system security to prevent weaknesses in the software system that can be exploited by an attacker. Description of software vulnerability is given in many different resources in different ways [61]. The most popular and widely utilized definition appears in the Common Vulnerabilities and Exposures (CVE) 2017 report as follows:

Vulnerability is a weakness in the computational logic found in software and some hardware components that, when exploited, results in a negative impact to confidentiality, integrity or availability.

Vulnerability analysis may require many different operations to identify defects and vulnerabilities in a software system. Vulnerabilities, which are a special kind of defect, are more critical than other defects because attackers exploit system vulnerabilities to perform unauthorized actions. A defect is a normal problem that can be encountered frequently in the system, easily found by users or developers and fixed promptly, whereas vulnerabilities are subtle mistakes in large codes [62, 63]. Wijayasekara et al. claim that some bugs have been identified as vulnerabilities after being publicly announced in bug databases [64]. These bugs are called "hidden impact vulnerabilities" or "hidden impact bugs." Therefore, the authors proposed a hidden impact vulnerability identification methodology that utilizes text mining techniques to determine which bugs in bug databases are vulnerabilities. According to the proposed method, a bug report was taken as input, and it produces feature vector after applying text mining. Then, classifier was applied and revealed whether it is a bug or a vulnerability. The results given in [64] demonstrate that a large proportion of discovered vulnerabilities were first described as hidden impact bugs in public bug databases. While bug reports were taken as input in that study, in many other studies, source code is taken as input. Text mining is a highly preferred technique for obtaining features directly from source codes as in the studies [65–69]. Several studies [63, 70] have compared text mining-based models and software metrics-based models.

In the security area of software systems, several studies have been conducted related to DM and ML. Some of these studies are compared in **Table 3**, which shows the data mining task and explanation of the studies, the year they were performed, the algorithms that were used, the type of vulnerability analysis, evaluation metrics, and results. In this table, the best performing algorithms according to the evaluation criteria are shown in bold.

Vulnerability analysis can be categorized into three types: static vulnerability analysis, dynamic vulnerability analysis, and hybrid analysis [61, 80]. Many studies have applied the static analysis approach, which detects vulnerabilities from source code without executing software, since it is cost-effective. Few studies have

Ref.	Year	Task	Objective	Algorithms	Type	Dataset description	Evaluation metrics and results
[71]	2011	Clustering	Obtaining software vulnerabilities based on RDBC	RDBC	Static	Database is built by RD-Entropy	FNR, FPR
[42]	2011	Classification/ regression	To predict the time to next vulnerability	LR, LMS, MLP, RBF, SMO	Static	NVD, CPE, CVSS	CC, RMSE, RRSE
[65]	2012	Text mining	Analysis of source code as text	RBF, SVM	Static	K9 email client for the Android platform	ACC, PR, recall ACC = 0.87, PR = 0.85, recall = 0.88
[64]	2012	Classification/ text mining	To identify vulnerabilities in bug databases	—	Static	Linux kernel MITRE CVE and MySQL bug databases	BDR, TPR, FPR 32% (Linux) and 62% (MySQL) of vulnerabilities
[72]	2014	Classification/ regression	Combine taint analysis and data mining to obtain vulnerabilities	ID3, C4.5/J48, RF, RT, KNN, NB, Bayes Net, MLP, SVM, LR	Hybrid	A version of WAP to collect the data	10-fold CV, TPD, ACC, PR, KAPPA ACC = 90.8%, PR = 92%, KAPPA = 81%
[73]	2014	Clustering	Identify vulnerabilities from source codes using CPG	—	Static	Neo4J and InfiniteGraph databases	—
[63]	2014	Classification	Comparison of software metrics with text mining	RF	Static	Vulnerabilities from open-source web apps (Drupal, Moodle, PHPMyAdmin)	3-fold CV, recall, IR, PR, FPR, ACC. Text mining provides benefits overall
[69]	2014	Classification	To create model in the form of a binary classifier using text mining	NB, RF	Static	Applications from the F-Droid repository and Android	10-fold CV, PR, recall PR and recall $\geq$ 80%
[74]	2015	Classification	A new approach (VCCFinder) to obtain potentially dangerous codes	SVM-based detection model	—	The database contains 66 GitHub projects	k-fold CV, false alarms <99% at the same level of recall
[70]	2015	Ranking/ classification	Comparison of text mining and software metrics models	RF	—	Vulnerabilities from open-source web apps (Drupal, Moodle, PHPMyAdmin)	10-fold CV Metrics: ER-BCE, ERBPP, ER-AVG
[75]	2015	Clustering	Search patterns for taint-style vulnerabilities in C code	Hierarchical clustering (complete-linkage)	Static	5 open-source projects: Linux, OpenSSL, Pidgin, VLC, Poppler (Xpdf)	Correct source, correct sanitization, number of traversals, generation time, execution time, reduction, amount of code review <95%

Ref.	Year	Task	Objective	Algorithms	Type	Dataset description	Evaluation metrics and results
[76]	2016	Classification	Static and dynamic features for classification	LR, MLP, RF	Hybrid	Dataset was created by analyzing 1039 test cases from the Debian Bug Tracker	FPR, FNR Detect 55% of vulnerable programs
[77]	2017	Classification	1. Employ a deep neural network 2. Combine N-gram analysis and feature selection	Deep neural network	—	Feature extraction from 4 applications (BoardGameGeek, Connectbot, CoolReader, AnkiDroid)	10 times using 5-fold CV ACC = 92.87%, PR = 94.71%, recall = 90.17%
[67]	2017	Text mining	To analyze characteristics of software vulnerability from source files	—	—	CVE, CWE, NVD databases	PR = 70%, recall = 60%
[68]	2017	Text mining	Deep learning (LSTM) is used to learn semantic and syntactic features in code	RNN, LSTM, DBN	—	Experiments on 18 Java applications from the Android OS platform	10-fold CV, PR, recall, and F-score Deep Belief Network PR, recall, and F-score > 80%
[66]	2018	Classification	Identify bugs by extracting text features from C source code	NB, KNN, K-means, NN, SVM, DT, RF	Static	NVD, Cat, Cp, Du, Echo, Head, Kill, Mkdir, Nl, Paste, Rm, Seq, Shuf, Sleep, Sort, Tail, Touch, Tr, Uniq, Wc, Whoami	5-fold CV ACC, TP, TN ACC = 74%
[78]	2018	Regression	A deep learning-based vulnerability detection system (VulDeePecker)	BLSTM NN	Static	NIST: NVD and SAR project	10-fold CV, PR, recall, F-score F-score = 80.8%
[79]	2018	Classification	A mapping between existing requirements and vulnerabilities	LR, SVM, NB	—	Data is gathered from Apache Tomcat, CVE, requirements from Bugzilla, and source code is collected from Github	PR, recall, F-score LSI > SVM

Table 3.
Data mining and machine learning studies on the subject "vulnerability analysis."

performed the dynamic analysis approach, in which one must execute software and check program behavior. The hybrid analysis approach [72, 76] combines these two approaches.

As revealed in **Table 3**, in addition to classification and text mining, clustering techniques are also frequently seen in software vulnerability analysis studies. To detect vulnerabilities in an unknown software data repository, entropy-based density clustering [71] and complete-linkage clustering [75] were proposed. Yamaguchi et al. [73] introduced a model to represent a large number of source codes as a graph called control flow graph (CPG), a combination of abstract syntax tree, CFG, and program dependency graph (PDG). This model enabled the discovery of previously unknown (zero-day) vulnerabilities.

To learn the time to next vulnerability, a prediction model was proposed in the study [42]. The result could be a number that refers to days or a bin representing values in a range. The authors used regression and classification techniques for the former and latter cases, respectively.

In vulnerability studies, issue tracking systems like Bugzilla, code repositories like Github, and vulnerability databases such as NVD, CVE, and CWE have been utilized [79]. In addition to these datasets, some studies have used Android [65, 68, 69] or web [63, 70, 72] (PHP source code) datasets. In recent years, researchers have concentrated on deep learning for building binary classifiers [77], obtaining vulnerability patterns [78], and learning long-term dependencies in sequential data [68] and features directly from the source code [81].

Li et al. [78] note two difficulties of vulnerability studies: demanding, intense manual labor and high false-negative rates. Thus, the widely used evaluation metrics in vulnerability analysis are false-positive rate and false-negative rate.

3.4 Data mining in design pattern mining

During the past years, software developers have used design patterns to create complex software systems. Thus, researchers have investigated the field of design patterns in many ways [82, 83]. Fowler defines a pattern as follows:

> "*A pattern is an idea that has been useful in one practical context and will probably be useful in others.*" [84]

Patterns display relationships and interactions between classes or objects. Well-designed object-oriented systems have various design patterns integrated into them. Design patterns can be highly useful for developers when they are used in the right manner and place. Thus, developers avoid recreating methods previously refined by others. The pattern approach was initially presented in 1994 by four authors—namely, Erich Gama, Richard Helm, Ralph Johnson, and John Vlissides—called the Gang of Four (GOF) in 1994 [85]. According to the authors, there are three types of design patterns:

1. Creational patterns provide an object creation mechanism to create the necessary objects based on predetermined conditions. They allow the system to call appropriate object and add flexibility to the system when objects are created. Some creational design patterns are factory method, abstract factory, builder, and singleton.

2. Structural patterns focus on the composition of classes and objects to allow the establishment of larger software groups. Some of the structural design patterns are adapter, bridge, composite, and decorator.

3. Behavioral patterns determine common communication patterns between objects and how multiple classes behave when performing a task. Some behavioral design patterns are command, interpreter, iterator, observer, and visitor.

Many design pattern studies exist in the literature. **Table 4** shows some design pattern mining studies related to machine learning and data mining. This table contains the aim of the study, mining task, year, and design patterns selected by the study, input data, dataset, and results of the studies.

In design pattern mining, detecting the design pattern is a frequent study objective. To do so, studies have used machine learning algorithms [87, 89–91], ensemble learning [95], deep learning [97], graph theory [94], and text mining [86, 95].

In study [91], the training dataset consists of 67 object-oriented (OO) metrics extracted by using the JBuilder tool. The authors used LRNN and decision tree techniques for pattern detection. Alhusain et al. [87] generated training datasets from existing pattern detection tools. The ANN algorithm was selected for pattern instances. Chihada et al. [90] created training data from pattern instances using 45 OO metrics. The authors utilized SVM for classifying patterns accurately. Another metrics-oriented dataset was developed by Dwivedi et al. [93]. To evaluate the results, the authors benefited from three open-source software systems (JHotDraw, QuickUML, and JUnit) and applied three classifiers, SVM, ANN, and RF. The advantage of using random forest is that it does not require linear features and can manage high-dimensional spaces.

To evaluate methods and to find patterns, open-source software projects such as JHotDraw, Junit, and MapperXML have been generally preferred by researchers. For example, Zanoni et al. [89] developed a tool called MARPLE-DPD by combining graph matching and machine learning techniques. Then, to obtain five design patterns, instances were collected from 10 open-source software projects, as shown in **Table 4**.

Design patterns and code smells are related issues: Code smell refers to symptoms in code, and if there are code smells in a software, its design pattern is not well constructed. Therefore, Kaur and Singh [96] checked whether design pattern and smell pairs appear together in a code by using J48 Decision Tree. Their obtained results showed that the singleton pattern had no presence of bad smells.

According to the studies summarized in the table, the most frequently used patterns are abstract factory and adapter. It has recently been observed that studies on ensemble learning in this field are increasing.

3.5 Data mining in refactoring

One of the SE tasks most often used to improve the quality of a software system is refactoring, which Martin Fowler has described as "a technique for restructuring an existing body of code, altering its internal structure without changing its external behavior" [98]. It improves readability and maintainability of the source code and decreases complexity of a software system. Some of the refactoring types are: Add Parameter, Replace Parameter, Extract method, and Inline method [99].

Code smell and refactoring are closely related to each other: Code smells represent problems due to bad design and can be fixed during refactoring. The main challenge is to obtain which part of the code needs refactoring.

Some of data mining studies related to software refactoring are presented in **Table 5**. Some studies focus on historical data to predict refactoring [100] or to obtain both refactoring and software defects [101] using different data mining algorithms such as LMT, Rip, and J48. Results suggest that when refactoring

Ref.	Year	Task	Objective	Algorithms	EL	Selected design patterns	Input data	Dataset	Evaluation metrics and results
[86]	2012	Text classification	Two-phase method: 1—text classification to 2—learning design patterns	NB, KNN, DT, SVM	—	46 security patterns, 34 Douglass patterns, 23 GoF patterns	Documents	Security, Douglass, GoF	PR, recall, EWM PR = 0.62, recall = 0.75
[87]	2013	Regression	An approach is to find a valid instance of a DP or not	ANN	—	Adapter, command, composite, decorator, observer, and proxy	Set of candidate classes	JHotDraw 5.1 open-source application	10 fold CV, PR, recall
[88]	2014	Graph mining	Sub-graph mining-based approach	CloseGraph	—	—	Java source code	Open-source project:YARI, Zest, JUnit, JFreeChart, ArgoUML	No any empirical comparison
[89]	2015	Classification/ clustering	MARPLE-DPD is developed to classify instances whether it is a bad or good instance	SVM, DT, RF, K-means, ZeroR, OneR, NB, JRip, CLOPE.	—	Classification for singleton and adapter Classification and clustering for composite, decorator, and factory method	—	10 open-source software systems DPExample, QuickUML 2001, Lexi v0.1.1 alpha, JRefactory v2.6.24, Netbeans v1.0.x, JUnit v3.7, JHotDraw v5.1, MapperXML v1.9.7, Nutch v0.4, PMD v1.8	10-fold CV, ACC, F-score, AUC ACC > =85%
[90]	2015	Regression	A new method (SVM-PHGS) is proposed	Simple Logistic, C4.5, KNN, SVM, SVM-PHGS	—	Adapter, builder, composite, factory method, iterator, observer	Source code	P-mart repository	PR, recall, F-score, FP PR = 0.81, recall =0.81, F-score = 0.81, FP = 0.038
[91]	2016	Classification	Design pattern recognition using ML algorithms.	LRNN, DT	—	Abstract factory, adapter patterns	Source code	Dataset with 67 OO metrics, extracted by JBuilder tool	5-fold CV, ACC, PR, recall, F-score ACC = 100% by LRNN

Ref.	Year	Task	Objective	Algorithms	EL	Selected design patterns	Input data	Dataset	Evaluation metrics and results
[92]	2016	Classification	Three aspects: design patterns, software metrics, and supervised learning methods	Layer Recurrent Neural Network (LRNN)	RF	Abstract factory, adapter, bridge, singleton, and template method	Source code	Dataset with 67 OO metrics, extracted by JBuilder tool	PR, recall, F-score F-score = 100% by LRNN and RF ACC = 100% by RF
[93]	2017	Classification	1. Creation of metrics-oriented dataset 2. Detection of software design patterns	ANN, SVM	RF	Abstract factory, adapter, bridge, composite, and Template	Source code	Metrics extracted from source codes (JHotDraw, QuickUML, and Junit)	5-fold and 10-fold CV, PR, recall, F-score ANN, SVM, and RF yielded to 100% PR for JHotDraw
[94]	2017	Classification	Detection of design motifs based on a set of directed semantic graphs	Strong graph simulation, graph matching	—	All three groups: creational, structural, behavioral	UML class diagrams	—	PR, recall High accuracy by the proposed method
[95]	2017	Text categorization	Selection of more appropriate design patterns	Fuzzy c-means	Ensemble-IG	Various design patterns	Problem definitions of design patterns	DP, GoF, Douglass, Security	F-score
[96]	2018	Classification	Finding design pattern and smell pairs which coexist in the code	J48	—	Used patterns: adapter, bridge, Template, singleton	Source code	Eclipse plugin Web of Patterns The tool selected for code smell detection is iPlasma	PR, recall, F-score, PRC, ROC Singleton pattern shows no presence of bad smells

Table 4.
Data mining and machine learning studies on the subject "design pattern mining."

Ref.	Year	Task	Objective	Algorithms	EL	Dataset	Evaluation metrics and results
[100]	2007	Regression	Stages: (1) data understanding, (2) preprocessing, (3) ML, (4) post-processing, (5) analysis of the results	J48, LMT, Rip, NNge	—	ArgoUML, Spring Framework	10-fold CV, PR, recall, F-score PR and recall are 0.8 for ArgoUML
[101]	2008	Classification	Finding the relationship between refactoring and defects	C4.5, LMT, Rip, NNge	—	ArgoUML, JBoss Cache, Liferay Portal, Spring Framework, XDoclet	PR, recall, F-score
[102]	2014	Regression	Propose GA-based learning for software refactoring based on ANN	GA, ANN	—	Xerces-J, JFreeChart, GanttProject, AntApache, JHotDraw, and Rhino.	Wilcoxon test with a 99% confidence level ($\alpha = 0.01$)
[103]	2015	Regression	Removing defects with time series in a multi-objective approach	Multi-objective algorithm, based on NSGA-II, ARIMA		FindBugs, JFreeChart, Hibernate, Pixelitor, and JDI-Ford	Wilcoxon rank sum test with a 99% confidence level ($\alpha < 1\%$)
[104]	2016	Web mining/ clustering	Unsupervised learning approach to detect refactoring opportunities in service-oriented applications	PAM, K-means, COBWEB, X-Means	—	Two datasets of WSDL documents	COBWEB and K-means max. 83.33% and 0%, inter-cluster COBWEB and K-means min. 33.33% and 66.66% intra-cluster
[105]	2017	Clustering	A novel algorithm (HASP) for software refactoring at the package level	Hierarchical clustering algorithm	—	Three open-source case studies	Modularization Quality and Evaluation Metric Function
[99]	2017	Classification	A technique to predict refactoring at class level	PCA, SMOTE LS-SVM, RBF	—	From tera- PROMISE Repository seven open-source software systems	10-fold CV, AUC, and ROC curves RBF kernel outperforms linear and polynomial kernel The mean value of AUC for LS-SVM RBF kernel is 0.96

Ref.	Year	Task	Objective	Algorithms	EL	Dataset	Evaluation metrics and results
[106]	2017	Classification	Exploring the impact of clone refactoring (CR) on the test code size	LR, KNN, NB	RF	data collected from an open-source Java software system (ANT)	PR, recall, accuracy, F-score kNN and RF outperform NB ACC (fitting (98%), LOOCV (95%), and 10 FCV (95%))
[107]	2017	—	Finding refactoring opportunities in source code	J48, BayesNet, SVM, LR	RF	Ant, ArgoUML, jEdit, jFreeChart, Mylyn	10-fold CV, PR, recall 86–97% PR and 71–98% recall for proposed tech
[108]	2018	Classification	A learning-based approach (CREC) to extract refactored and non-refactored clone groups from repositories	C4.5, SMO, NB.	RF, Adaboost	Axis2, Eclipse.jdt.core, Elastic Search, JFreeChart, JRuby, and Lucene	PR, recall, F-score F-score = 83% in the within-project F-score = 76% in the cross-project
[109]	2018	Clustering	Combination of the use of multi-objective and unsupervised learning to decrease developer's effort	GMM, EM	—	ArgoUML, JHotDraw, GanttProject, UTest, Apache Ant, Azureus	One-way ANOVA with a 95% confidence level (α = 5%)

Table 5.
Data mining and machine learning studies on the subject "refactoring."

increases, the number of software defects decreases, and thus refactoring has a positive effect on software quality.

While automated refactoring does not always give the desired result, manual refactoring is time-consuming. Therefore, one study [109] proposed a clustering-based recommendation tool by combining multi-objective search and unsupervised learning algorithm to reduce the number of refactoring options. At the same time, the number of refactoring that should be selected is decreasing with the help of the developer's feedback.

4. Discussion

Since many SE studies that apply data mining approaches exist in the literature, this article presents only a few of them. However, **Figure 4** shows the current number of papers obtained from the Scopus search engine for each year from 2010 to 2019 by using queries in the title/abstract/keywords field. We extracted publications in 2020 since this year has not completed yet. Queries included ("data mining" OR "machine learning") with ("defect prediction" OR "defect detection" OR "bug prediction" OR "bug detection") for defect prediction, ("effort estimation" OR "effort prediction" OR "cost estimation") for effort estimation, ("vulnerab*" AND "software" OR "vulnerability analysis") for vulnerability analysis, and ("software" AND "refactoring") for refactoring. As seen in the figure, the number of studies using data mining in SE tasks, especially defect prediction and vulnerability analysis, has increased rapidly. The most stable area in the studies is design pattern mining.

Figure 5 shows the publications studied in classification, clustering, text mining, and association rule mining as a percentage of the total number of papers obtained by a Scopus query for each SE task. For example, in defect prediction, the number of studies is 339 in the field of classification, 64 in clustering, 8 in text mining, and 25 in the field of association rule mining. As can be seen from the pie charts, while clustering is a popular DM technique in refactoring, no study related to text mining is found in this field. In other SE tasks, the preferred technique is classification, and the second is clustering.

Defect prediction generally compares learning algorithms in terms of whether they find defects correctly using classification algorithms. Besides this approach, in some studies, clustering algorithms were used to select futures [110] or to compare supervised and unsupervised methods [27]. In the text mining area, to extract features from scripts, TF-IDF techniques were generally used [111, 112]. Although many different algorithms have been used in defect prediction, the most popular ones are NB, MLP, and RBF.

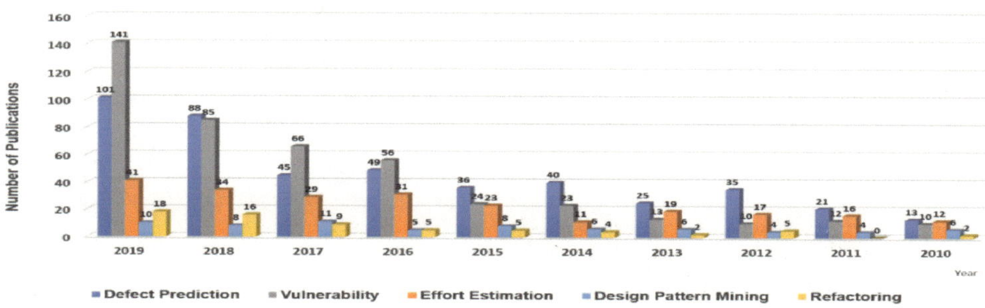

Figure 4.
Number of publications of data mining studies for SE tasks from Scopus search by their years.

Figure 6 shows the number of document types (conference paper, book chapter, article, book) published between the years of 2010 and 2019. It is clearly seen that conference papers and articles are the most preferred research study type. It is clearly seen that there is no review article about data mining studies in design pattern mining.

Table 6 shows popular repositories that contain various datasets and their descriptions, which tasks they are used for, and hyperlinks to download. For

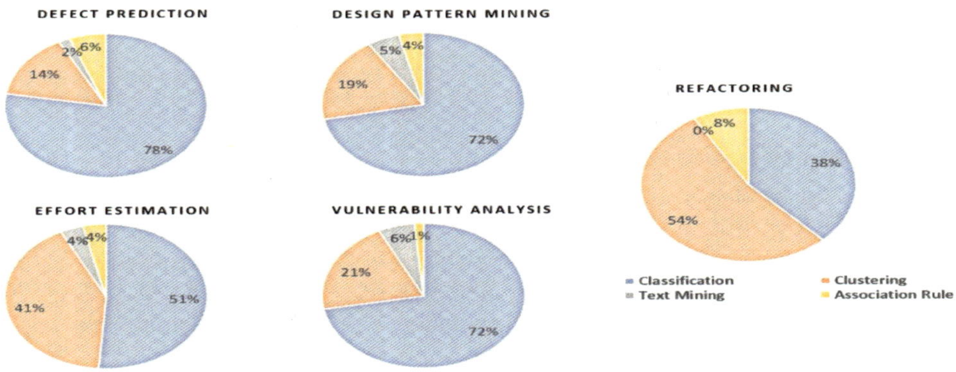

Figure 5.
Number of publications of data mining studies for SE tasks from Scopus search by their topics.

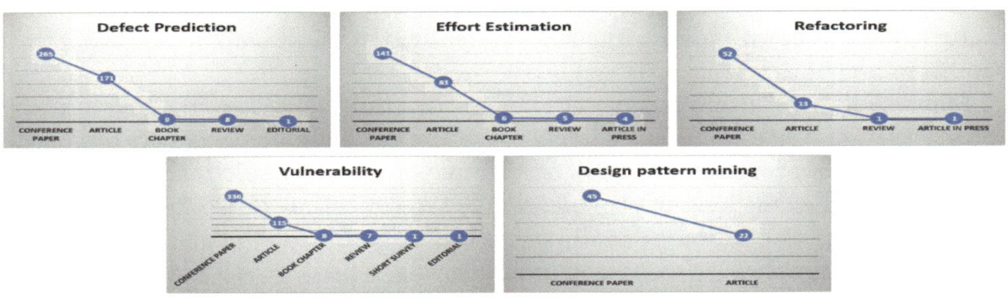

Figure 6.
The number of publications in terms of document type between 2010 and 2019.

Repository	Topic	Description	Web link
Nasa MDP	Defect Pred.	NASA's Metrics Data Program	https://github.com/opensciences/opensciences.github.io/tree/master/repo/defect/mccabehalsted/_posts
Android Git	Defect Pred.	Android version bug reports	https://android.googlesource.com/
PROMISE	Defect Pred. Effort Est.	It includes 20 datasets for defect prediction and cost estimation	http://promise.site.uottawa.ca/SERepository/datasets-page.html
Software Defect Pred. Data	Defect Pred.	It includes software metrics, # of defects, etc. Eclipse JDT: Eclipse PDE:	http://www.seiplab.riteh.uniri.hr/?page_id=834&lang=en
PMART	Design pattern mining	It has 22 patterns 9 Projects, 139 ins. Format: XML Manually detected and validated	http://www.ptidej.net/tools/designpatterns/

Table 6.
Description of popular repositories used in studies.

example, the PMART repository includes source files of java projects, and the PROMISE repository has different datasets with software metrics such as cyclomatic complexity, design complexity, and lines of code. Since these repositories contain many datasets, no detailed information about them has been provided in this article.

Refactoring can be applied at different levels; study [105] predicted refactoring at package level using hierarchical clustering, and another study [99] applied class-level refactoring using LS-SVM as learning algorithm, SMOTE for handling refactoring, and PCA for feature extraction.

5. Conclusion and future work

Data mining techniques have been applied successfully in many different domains. In software engineering, to improve the quality of a product, it is highly critical to find existing deficits such as bugs, defects, code smells, and vulnerabilities in the early phases of SDLC. Therefore, many data mining studies in the past decade have aimed to deal with such problems. The present paper aims to provide information about previous studies in the field of software engineering. This survey shows how classification, clustering, text mining, and association rule mining can be applied in five SE tasks: defect prediction, effort estimation, vulnerability analysis, design pattern mining, and refactoring. It clearly shows that classification is the most used DM technique. Therefore, new studies can focus on clustering on SE tasks.

Abbreviations

LMT	logistic model trees
Rip	repeated incremental pruning
NNge	nearest neighbor generalization
PCA	principal component analysis
PAM	partitioning around medoids
LS-SVM	least-squares support vector machines
MAE	mean absolute error
RBF	radial basis function
RUS	random undersampling
SMO	sequential minimal optimization
GMM	Gaussian mixture model
EM	expectation maximizaion
LR	logistic regression
SMB	SMOTEBoost
RUS-bal	balanced version of random undersampling
THM	threshold-moving
BNC	AdaBoost.NC
RF	random forest
RBF	radial basis function
CC	correlation coefficient
ROC	receiver operating characteristic
BayesNet	Bayesian network
SMOTE	synthetic minority over-sampling technique

Author details

Elife Ozturk Kiyak
Graduate School of Natural and Applied Sciences, Dokuz Eylul University, Turkey

*Address all correspondence to: elife.ozturk@ceng.deu.edu.tr

References

[1] Halkidi M, Spinellis D, Tsatsaronis G, Vazirgiannis M. Data mining in software engineering. Intelligent Data Analysis. 2011;**15**(3):413-441. DOI: 10.3233/IDA-2010-0475

[2] Dhamija A, Sikka S. A review paper on software engineering areas implementing data mining tools & techniques. International Journal of Computational Intelligence Research. 2017;**13**(4):559-574

[3] Minku LL, Mendes E, Turhan B. Data mining for software engineering and humans in the loop. Progress in Artificial Intelligence. 2016;**5**(4): 307-314

[4] Malhotra R. A systematic review of machine learning techniques for software fault prediction. Applied Soft Computing. 2015;**27**:504-518. DOI: 10.1016/j.asoc.2014.11.023

[5] Mayvan BB, Rasoolzadegan A, Ghavidel Yazdi Z. The state of the art on design patterns: A systematic mapping of the literature. Journal of Systems and Software. 2017;**125**:93-118. DOI: 10.1016/j.jss.2016.11.030

[6] Sehra SK, Brar YS, Kaur N, Sehra SS. Research patterns and trends in software effort estimation. Information and Software Technology. 2017;**91**:1-21. DOI: 10.1016/j.infsof.2017.06.002

[7] Taylor Q, Giraud-Carrier C, Knutson CD. Applications of data mining in software engineering. International Journal of Data Analysis Techniques and Strategies. 2010;**2**(3):243-257

[8] Coelho RA, Guimarães FRN, Esmin AA. Applying swarm ensemble clustering technique for fault prediction using software metrics. In: Machine Learning and Applications (ICMLA), 2014 13th International Conference on IEEE. 2014. pp. 356-361

[9] Prasad MC, Florence L, Arya A. A study on software metrics based software defect prediction using data mining and machine learning techniques. International Journal of Database Theory and Application. 2015;**8**(3):179-190. DOI: 10.14257/ijdta.2015.8.3.15

[10] Zhang Y, Lo D, Xia X, Sun J. Combined classifier for cross-project defect prediction: An extended empirical study. Frontiers of Computer Science. 2018;**12**(2):280-296. DOI: 10.1007/s11704-017-6015-y

[11] Yang X, Lo D, Xia X, Zhang Y, Sun J. Deep learning for just-in-time defect prediction. In: International Conference on Software Quality, Reliability and Security (QRS); 3–5 August 2015; Vancouver, Canada: IEEE; 2015. pp. 17-26

[12] Zhang F, Zheng Q, Zou Y, Hassan AE. Cross-project defect prediction using a connectivity-based unsupervised classifier. In: Proceedings of the 38th International Conference on Software Engineering ACM; 14–22 May 2016; Austin, TX, USA: IEEE; 2016. pp. 309-320

[13] Di Nucci D, Palomba F, Oliveto R, De Lucia A. Dynamic selection of classifiers in bug prediction: An adaptive method. IEEE Transactions on Emerging Topics in Computational Intelligence. 2017;**1**(3):202-212. DOI: 10.1109/TETCI.2017.2699224

[14] Zimmermann T, Nagappan N, Gall H, Giger E, Murphy B. Cross-project defect prediction: A large scale experiment on data vs. domain vs. process. In: Proceedings of the 7th Joint Meeting of the European Software Engineering Conference and the Symposium on the Foundations of Software Engineering (ESEC/FSE '09); August 2009; Amsterdam, Netherlands: ACM; 2009. pp. 91-100

[15] Turhan B, Menzies T, Bener AB, Di Stefano J. On the relative value of cross-company and within-company data for defect prediction. Empirical Software Engineering. 2009;**14**(5):540-578. DOI: 10.1007/s10664-008-9103-7

[16] Herbold S, Trautsch A, Grabowski J. A comparative study to benchmark cross-project defect prediction approaches. IEEE Transactions on Software Engineering. 2017;**44**(9): 811-833. DOI: 10.1109/TSE.2017.2724538

[17] Ghotra B, McIntosh S, Hassan AE. Revisiting the impact of classification techniques on the performance of defect prediction models. In: IEEE/ACM 37th IEEE International Conference on Software Engineering; 16–24 May 2015; Florence, Italy: IEEE; 2015. pp. 789-800

[18] Wang T, Li W, Shi H, Liu Z. Software defect prediction based on classifiers ensemble. Journal of Information & Computational Science. 2011;**8**:4241-4254

[19] Wang S, Yao X. Using class imbalance learning for software defect prediction. IEEE Transactions on Reliability. 2013;**62**:434-443. DOI: 10.1109/TR.2013.2259203

[20] Rodriguez D, Herraiz I, Harrison R, Dolado J, Riquelme JC. Preliminary comparison of techniques for dealing with imbalance in software defect prediction. In: Proceedings of the 18th International Conference on Evaluation and Assessment in Software Engineering; May 2014; London, United Kingdom: ACM; 2014. p. 43

[21] Laradji IH, Alshayeb M, Ghouti L. Software defect prediction using ensemble learning on selected features. Information and Software Technology. 2015;**58**:388-402. DOI: 10.1016/j.infsof.2014.07.005

[22] Malhotra R, Raje R. An empirical comparison of machine learning techniques for software defect prediction. In: Proceedings of the 8th International Conference on Bioinspired Information and Communications Technologies. Boston, Massachusetts; December 2014. pp. 320-327

[23] Malhotra R. An empirical framework for defect prediction using machine learning techniques with Android software. Applied Soft Computing. 2016;**49**:1034-1050. DOI: 10.1016/j.asoc.2016.04.032

[24] Tantithamthavorn C, McIntosh S, Hassan AE, Matsumoto K. Automated parameter optimization of classification techniques for defect prediction models. In: Proceedings of the 38th International Conference on Software Engineering (ICSE '16). Austin, Texas; May 2016. pp. 321-332

[25] Kumar L, Misra S, Rath SK. An empirical analysis of the effectiveness of software metrics and fault prediction model for identifying faulty classes. Computer Standards & Interfaces. 2017; **53**:1-32. DOI: 10.1016/j.csi.2017.02.003

[26] Yang X, Lo D, Xia X, Sun J. TLEL: A two-layer ensemble learning approach for just-in-time defect prediction. Information and Software Technology. 2017;**87**:206-220. DOI: 10.1016/j.infsof.2017.03.007

[27] Chen X, Zhao Y, Wang Q, Yuan Z. MULTI: Multi-objective effort-aware just-in-time software defect prediction. Information and Software Technology. 2018;**93**:1-13. DOI: 10.1016/j.infsof.2017.08.004

[28] Zimmermann T, Premraj R, Zeller A. Predicting defects for eclipse. In: Third International Workshop on Predictor Models in Software Engineering (PROMISE'07); 20-26 May 2007; Minneapolis, USA: IEEE; 2007. p. 9

[29] Prakash VA, Ashoka DV, Aradya VM. Application of data mining

techniques for defect detection and classification. In: Proceedings of the 3rd International Conference on Frontiers of Intelligent Computing: Theory and Applications (FICTA); 14–15 November 2014; Odisha, India; 2014. pp. 387-395

[30] Yousef AH. Extracting software static defect models using data mining. Ain Shams Engineering Journal. 2015;**6**: 133-144. DOI: 10.1016/j.asej.2014.09.007

[31] Gupta DL, Saxena K. AUC based software defect prediction for object-oriented systems. International Journal of Current Engineering and Technology. 2016;**6**:1728-1733

[32] Kumar L, Rath SK. Application of genetic algorithm as feature selection technique in development of effective fault prediction model. In: IEEE Uttar Pradesh Section International Conference on Electrical, Computer and Electronics Engineering (UPCON); 9-11 December 2016; Varanasi, India: IEEE; 2016. pp. 432-437

[33] Tomar D, Agarwal S. Prediction of defective software modules using class imbalance learning. Applied Computational Intelligence and Soft Computing. 2016;**2016**:1-12. DOI: 10.1155/2016/7658207

[34] Ryu D, Baik J. Effective multi-objective naïve Bayes learning for cross-project defect prediction. Applied Soft Computing. 2016;**49**:1062-1077. DOI: 10.1016/j.asoc.2016.04.009

[35] Ali MM, Huda S, Abawajy J, Alyahya S, Al-Dossari H, Yearwood J. A parallel framework for Software Defect detection and metric selection on cloud computing. Cluster Computing. 2017; **20**:2267-2281. DOI: 10.1007/s10586-017-0892-6

[36] Wijaya A, Wahono RS. Tackling imbalanced class in software defect prediction using two-step cluster based random undersampling and stacking technique. Jurnal Teknologi. 2017;**79**: 45-50

[37] Singh PD, Chug A. Software defect prediction analysis using machine learning algorithms. In: 7th International Conference on Cloud Computing, Data Science & Engineering-Confluence; 2–13 January 2017; Noida, India: IEEE; 2017. pp. 775-781

[38] Hammouri A, Hammad M, Alnabhan M, Alsarayrah F. Software bug prediction on using machine learning approach. International Journal of Advanced Computer Science and Applications. 2018;**9**:78-83

[39] Akour M, Alsmadi I, Alazzam I. Software fault proneness prediction: A comparative study between bagging, boosting, and stacking ensemble and base learner methods. International Journal of Data Analysis Techniques and Strategies. 2017;**9**:1-16

[40] Bowes D, Hall T, Petric J. Software defect prediction: Do different classifiers find the same defects? Software Quality Journal. 2018;**26**: 525-552. DOI: 10.1007/s11219-016-9353-3

[41] Watanabe T, Monden A, Kamei Y, Morisaki S. Identifying recurring association rules in software defect prediction. In: IEEE/ACIS 15th International Conference on Computer and Information Science (ICIS); 26–29 June 2016; Okayama, Japan: IEEE; 2016. pp. 1-6

[42] Zhang S, Caragea D, Ou X. An empirical study on using the national vulnerability database to predict software vulnerabilities. In: International Conference on Database and Expert Systems Applications. Berlin, Heidelberg: Springer; 2011. pp. 217-223

[43] Wen J, Li S, Lin Z, Hu Y, Huang C. Systematic literature review of machine

learning based software development effort estimation models. Information and Software Technology. 2012;**54**: 41-59. DOI: 10.1016/j. infsof.2011.09.002

[44] Dave VS, Dutta K. Neural network based models for software effort estimation: A review. Artificial Intelligence Review. 2014;**42**:295-307. DOI: 10.1007/s10462-012-9339-x

[45] Kultur Y, Turhan B, Bener AB. ENNA: Software effort estimation using ensemble of neural networks with associative memory. In: Proceedings of the 16th ACM SIGSOFT; November 2008; Atlanta, Georgia: ACM; 2008. pp. 330-338

[46] Kultur Y, Turhan B, Bener A. Ensemble of neural networks with associative memory (ENNA) for estimating software development costs. Knowledge-Based Systems. 2009;**22**: 395-402. DOI: 10.1016/j. knosys.2009.05.001

[47] Corazza A, Di Martino S, Ferrucci F, Gravino C, Mendes E. Investigating the use of support vector regression for web effort estimation. Empirical Software Engineering. 2011;**16**:211-243. DOI: 10.1007/s10664-010-9138-4

[48] Minku LL, Yao X. A principled evaluation of ensembles of learning machines for software effort estimation. In: Proceedings of the 7th International Conference on Predictive Models in Software Engineering; September 2011; Banff, Alberta, Canada: ACM; 2011. pp. 1-10

[49] Minku LL, Yao X. Software effort estimation as a multiobjective learning problem. ACM Transactions on Software Engineering and Methodology (TOSEM). 2013;**22**:35. DOI: 10.1145/ 2522920.2522928

[50] Minku LL, Yao X. Can cross-company data improve performance in software effort estimation? In: Proceedings of the 8th International Conference on Predictive Models in Software Engineering (PROMISE '12); September 2012; New York, United States: ACM; 2012. pp. 69-78

[51] Kocaguneli E, Menzies T, Keung JW. On the value of ensemble effort estimation. IEEE Transactions on Software Engineering. 2012;**38**: 1403-1416. DOI: 10.1109/TSE.2011.111

[52] Dejaeger K, Verbeke W, Martens D, Baesens B. Data mining techniques for software effort estimation. IEEE Transactions on Software Engineering. 2011;**38**:375-397. DOI: 10.1109/ TSE.2011.55

[53] Khatibi V, Jawawi DN, Khatibi E. Increasing the accuracy of analogy based software development effort estimation using neural networks. International Journal of Computer and Communication Engineering. 2013;**2**:78

[54] Subitsha P, Rajan JK. Artificial neural network models for software effort estimation. International Journal of Technology Enhancements and Emerging Engineering Research. 2014;**2**:76-80

[55] Maleki I, Ghaffari A, Masdari M. A new approach for software cost estimation with hybrid genetic algorithm and ant colony optimization. International Journal of Innovation and Applied Studies. 2014;**5**:72

[56] Huang J, Li YF, Xie M. An empirical analysis of data preprocessing for machine learning-based software cost estimation. Information and Software Technology. 2015;**67**:108-127. DOI: 10.1016/j.infsof.2015.07.004

[57] Nassif AB, Azzeh M, Capretz LF, Ho D. Neural network models for software development effort estimation. Neural Computing and Applications. 2016;**27**:2369-2381. DOI: 10.1007/ s00521-015-2127-1

[58] Zare F, Zare HK, Fallahnezhad MS. Software effort estimation based on the optimal Bayesian belief network. Applied Soft Computing. 2016;**49**:968-980. DOI: 10.1016/j.asoc.2016.08.004

[59] Azzeh M, Nassif AB. A hybrid model for estimating software project effort from use case points. Applied Soft Computing. 2016;**49**:981-989. DOI: 10.1016/j.asoc.2016.05.008

[60] Hidmi O, Sakar BE. Software development effort estimation using ensemble machine learning. International Journal of Computing, Communication and Instrumentation Engineering. 2017;**4**:143-147

[61] Ghaffarian SM, Shahriari HR. Software vulnerability analysis and discovery using machine-learning and data-mining techniques. ACM Computing Surveys (CSUR). 2017;**50**: 1-36. DOI: 10.1145/3092566

[62] Jimenez M, Papadakis M, Le Traon Y. Vulnerability prediction models: A case study on the linux kernel. In: IEEE 16th International Working Conference on Source Code Analysis and Manipulation (SCAM); 2–3 October 2016; Raleigh, NC, USA: IEEE; 2016. pp. 1-10

[63] Walden J, Stuckman J, Scandariato R. Predicting vulnerable components: Software metrics vs text mining. In: IEEE 25th International Symposium on Software Reliability Engineering; 3–6 November 2014; Naples, Italy: IEEE; 2014. pp. 23-33

[64] Wijayasekara D, Manic M, Wright JL, McQueen M. Mining bug databases for unidentified software vulnerabilities. In: 5th International Conference on Human System Interactions; 6–8 June 2012; Perth, WA, Australia: IEEE; 2013. pp. 89-96

[65] Hovsepyan A, Scandariato R, Joosen W, Walden J. Software vulnerability prediction using text analysis techniques. In: Proceedings of the 4th International Workshop on Security Measurements and Metrics (ESEM '12); September 2012; Lund Sweden: IEEE; 2012. pp. 7-10

[66] Chernis B, Verma R. Machine learning methods for software vulnerability detection. In: Proceedings of the Fourth ACM International Workshop on Security and Privacy Analytics (CODASPY '18); March 2018; Tempe, AZ, USA: 2018. pp. 31-39

[67] Li X, Chen J, Lin Z, Zhang L, Wang Z, Zhou M, et al. Mining approach to obtain the software vulnerability characteristics. In: 2017 Fifth International Conference on Advanced Cloud and Big Data (CBD); 13–16 August 2017; Shanghai, China: IEEE; 2017. pp. 296-301

[68] Dam HK, Tran T, Pham T, Ng SW, Grundy J, Ghose A. Automatic feature learning for vulnerability prediction. arXiv preprint arXiv:170802368 2017

[69] Scandariato R, Walden J, Hovsepyan A, Joosen W. Predicting vulnerable software components via text mining. IEEE Transactions on Software Engineering. 2014;**40**: 993-1006

[70] Tang Y, Zhao F, Yang Y, Lu H, Zhou Y, Xu B. Predicting vulnerable components via text mining or software metrics? An effort-aware perspective. In: IEEE International Conference on Software Quality, Reliability and Security; 3–5 August 2015; Vancouver, BC, Canada: IEEE; 2015. p. 27–36

[71] Wang Y, Wang Y, Ren J. Software vulnerabilities detection using rapid density-based clustering. Journal of Information and Computing Science. 2011;**8**:3295-3302

[72] Medeiros I, Neves NF, Correia M. Automatic detection and correction of

web application vulnerabilities using data mining to predict false positives. In: Proceedings of the 23rd International Conference on World Wide Web (WWW '14); April 2014; Seoul, Korea; 2014. pp. 63-74

[73] Yamaguchi F, Golde N, Arp D, Rieck K. Modeling and discovering vulnerabilities with code property graphs. In: 2014 IEEE Symposium on Security and Privacy; 18-21 May 2014; San Jose, CA, USA: IEEE; 2014. pp. 590-604

[74] Perl H, Dechand S, Smith M, Arp D, Yamaguchi F, Rieck K, et al. Vccfinder: Finding Potential Vulnerabilities in Open-source Projects to Assist Code Audits. In: 22nd ACM Conference on Computer and Communications Security (CCS'15). Denver, Colorado, USA; 2015. pp. 426-437

[75] Yamaguchi F, Maier A, Gascon H, Rieck K. Automatic inference of search patterns for taint-style vulnerabilities. In: 2015 IEEE Symposium on Security and Privacy; San Jose, CA, USA: IEEE; 2015. pp. 797-812

[76] Grieco G, Grinblat GL, Uzal L, Rawat S, Feist J, Mounier L. Toward large-scale vulnerability discovery using machine learning. In: Proceedings of the Sixth ACM Conference on Data and Application Security and Privacy; March 2016; New Orleans, Louisiana, USA; 2016. pp. 85-96

[77] Pang Y, Xue X, Wang H. Predicting vulnerable software components through deep neural network. In: Proceedings of the 2017 International Conference on Deep Learning Technologies; June 2017; Chengdu, China; 2017. pp. 6-10

[78] Li Z, Zou D, Xu S, Ou X, Jin H, Wang S, et al. VulDeePecker: A Deep Learning-Based System for Vulnerability Detection. arXiv preprint arXiv:180101681. 2018

[79] Imtiaz SM, Bhowmik T. Towards data-driven vulnerability prediction for requirements. In: Proceedings of the 2018 26th ACM Joint Meeting on European Software Engineering Conference and Symposium on the Foundations of Software Engineering; November, 2018; Lake Buena Vista, FL, USA. 2018. pp. 744-748

[80] Jie G, Xiao-Hui K, Qiang L. Survey on software vulnerability analysis method based on machine learning. In: IEEE First International Conference on Data Science in Cyberspace (DSC); 13–16 June 2016; Changsha, China: IEEE; 2017. pp. 642-647

[81] Russell R, Kim L, Hamilton L, Lazovich T, Harer J, Ozdemir O, et al. Automated vulnerability detection in source code using deep representation learning. In: 17th IEEE International Conference on Machine Learning and Applications (ICMLA). Orlando, FL, USA: IEEE; 2018, 2019. pp. 757-762

[82] Mayvan BB, Rasoolzadegan A, Yazdi ZG. The state of the art on design patterns: A systematic mapping of the literature. Journal of Systems and Software. 2017;**125**:93-118. DOI: 10.1016/j.jss.2016.11.030

[83] Dong J, Zhao Y, Peng T. A review of design pattern mining techniques. International Journal of Software Engineering and Knowledge Engineering. 2009;**19**:823-855. DOI: 10.1142/S021819400900443X

[84] Fowler M. Analysis Patterns: Reusable Object Models. Boston: Addison-Wesley Professional; 1997

[85] Vlissides J, Johnson R, Gamma E, Helm R. Design Patterns-Elements of Reusable Object-Oriented Software. 1st ed. Addison-Wesley Professional; 1994

[86] Hasheminejad SMH, Jalili S. Design patterns selection: An automatic two-phase method. Journal of Systems and

Software. 2012;**85**:408-424. DOI: 10.1016/j.jss.2011.08.031

[87] Alhusain S, Coupland S, John R, Kavanagh M. Towards machine learning based design pattern recognition. In: 2013 13th UK Workshop on Computational Intelligence (UKCI); 9–11 September 2013; Guildford, UK: IEEE; 2013. pp. 244-251

[88] Tekin U. Buzluca F, A graph mining approach for detecting identical design structures in object-oriented design models. Science of Computer Programming. 2014;**95**:406-425. DOI: 10.1016/j.scico.2013.09.015

[89] Zanoni M, Fontana FA, Stella F. On applying machine learning techniques for design pattern detection. Journal of Systems and Software. 2015;**103**: 102-117. DOI: 10.1016/j.jss.2015.01.037

[90] Chihada A, Jalili S, Hasheminejad SMH, Zangooei MH. Source code and design conformance, design pattern detection from source code by classification approach. Applied Soft Computing. 2015;**26**:357-367. DOI: 10.1016/j.asoc.2014.10.027

[91] Dwivedi AK, Tirkey A, Ray RB, Rath SK. Software design pattern recognition using machine learning techniques. In: 2016 IEEE Region 10 Conference (TENCON); 22–25 November 2016; Singapore, Singapore: IEEE; 2017. pp. 222-227

[92] Dwivedi AK, Tirkey A, Rath SK. Applying software metrics for the mining of design pattern. In: IEEE Uttar Pradesh Section International Conference on Electrical, Computer and Electronics Engineering (UPCON); 9–11 December 2016; Varanasi, India: IEEE; 2017. pp. 426-431

[93] Dwivedi AK, Tirkey A, Rath SK. Software design pattern mining using classification-based techniques. Frontiers of Computer Science. 2018;**12**:908-922. DOI: 10.1007/s11704-017-6424-y

[94] Mayvan BB, Rasoolzadegan A. Design pattern detection based on the graph theory. Knowledge-Based Systems. 2017;**120**:211-225. DOI: 10.1016/j.knosys.2017.01.007

[95] Hussain S, Keung J, Khan AA. Software design patterns classification and selection using text categorization approach. Applied Soft Computing. 2017;**58**:225-244. DOI: 10.1016/j.asoc.2017.04.043

[96] Kaur A, Singh S. Detecting software bad smells from software design patterns using machine learning algorithms. International Journal of Applied Engineering Research. 2018;**13**: 10005-10010

[97] Hussain S, Keung J, Khan AA, Ahmad A, Cuomo S, Piccialli F. Implications of deep learning for the automation of design patterns organization. Journal of Parallel and Distributed Computing. 2018;**117**: 256-266. DOI: 10.1016/j.jpdc.2017.06.022

[98] Fowler M. Refactoring: Improving the Design of Existing Code. 2nd ed. Boston: Addison-Wesley Professional; 2018

[99] Kumar L, Sureka A. Application of LSSVM and SMOTE on seven open source projects for predicting refactoring at class level. In: 24th Asia-Pacific Software Engineering Conference (APSEC); 4–8 December 2017; Nanjing, China: IEEE; 2018. pp. 90-99

[100] Ratzinger J, Sigmund T, Vorburger P, Gall H. Mining software evolution to predict refactoring. In: First International Symposium on Empirical Software Engineering and Measurement (ESEM 2007); 20–21 September 2007; Madrid, Spain: IEEE; 2007. pp. 354-363

[101] Ratzinger J, Sigmund T, Gall HC. On the relation of refactoring and

software defects. In: Proceedings of the 2008 International Working Conference on Mining Software Repositories; May 2008; Leipzig, Germany: ACM; 2008. pp. 35-38

[102] Amal B, Kessentini M, Bechikh S, Dea J, Said LB. On the Use of Machine Learning and Search-Based software engineering for ill-defined fitness function: A case study on software refactoring. In: International Symposium on Search Based Software Engineering; 26-29 August 2014; Fortaleza, Brazil; 2014. pp. 31-45

[103] Wang H, Kessentini M, Grosky W, Meddeb H. On the use of time series and search based software engineering for refactoring recommendation. In: Proceedings of the 7th International Conference on Management of Computational and Collective intElligence in Digital EcoSystems. Caraguatatuba, Brazil; October 2015. pp. 35-42

[104] Rodríguez G, Soria Á, Teyseyre A, Berdun L, Campo M. Unsupervised learning for detecting refactoring opportunities in service-oriented applications. In: International Conference on Database and Expert Systems Applications; 5–8 September; Porto, Portugal: Springer; 2016. pp. 335-342

[105] Marian Z, Czibula IG, Czibula G. A hierarchical clustering-based approach for software restructuring at the package level. In: 9th International Symposium on Symbolic and Numeric Algorithms for Scientific Computing (SYNASC); 21–24 September 2017; Timisoara, Romania: IEEE; 2018. pp. 239-246

[106] Mourad B, Badri L, Hachemane O, Ouellet A. Exploring the impact of clone refactoring on test code size in object-oriented software. In: 16th IEEE International Conference on Machine Learning and Applications (ICMLA); 18-21 December 2017; Cancun, Mexico. 2018. pp. 586-592

[107] Imazato A, Higo Y, Hotta K, Kusumoto S. Finding extract method refactoring opportunities by analyzing development history. In: IEEE 41st Annual Computer Software and Applications Conference (COMPSAC); 4–8 July 2017; Turin, Italy: IEEE; 2018. pp. 190-195

[108] Yue R, Gao Z, Meng N, Xiong Y, Wang X. Automatic clone recommendation for refactoring based on the present and the past. In: IEEE International Conference on Software Maintenance and Evolution (ICSME); 23–29 September 2018; Madrid, Spain: IEEE; 2018. pp. 115-126

[109] Alizadeh V, Kessentini M. Reducing interactive refactoring effort via clustering-based multi-objective search. In: 33rd ACM/IEEE International Conference on Automated Software Engineering; September 2018; Montpellier, France: ACM/IEEE; 2018. pp. 464-474

[110] Ni C, Liu WS, Chen X, Gu Q, Chen DX, Huang QG. A cluster based feature selection method for cross-project software defect prediction. Journal of Computer Science and Technology. 2017;**32**:1090-1107. DOI: 10.1007/s11390-017-1785-0

[111] Rahman A, Williams L. Characterizing defective configuration scripts used for continuous deployment. In: 11th International Conference on Software Testing, Verification and Validation (ICST); 9–13 April 2018; Vasteras, Sweden: IEEE; 2018. pp. 34-45

[112] Kukkar A, Mohana R. A supervised bug report classification with incorporate and textual field knowledge. Procedia Computer Science. 2018;**132**: 352-361. DOI: 10.1016/j.procs.2018.05.194

Chapter 9

Data Mining for Student Performance Prediction in Education

Ferda Ünal

Abstract

The ability to predict the performance tendency of students is very important to improve their teaching skills. It has become a valuable knowledge that can be used for different purposes; for example, a strategic plan can be applied for the development of a quality education. This paper proposes the application of data mining techniques to predict the final grades of students based on their historical data. In the experimental studies, three well-known data mining techniques (decision tree, random forest, and naive Bayes) were employed on two educational datasets related to mathematics lesson and Portuguese language lesson. The results showed the effectiveness of data mining learning techniques when predicting the performances of students.

Keywords: data mining, student performance prediction, classification

1. Introduction

Recently, online systems in education have increased, and student digital data has come to big data size. This makes possible to draw rules and predictions about the students by processing educational data with data mining techniques. All kinds of information about the student's socioeconomic environment, learning environment, or course notes can be used for prediction, which affect the success or failure of a student.

In this study, the successes of the students at the end of the semester are estimated by using the student data obtained from secondary education of two Portuguese schools. The aim of this study is to predict the students' final grades to support the educators to take precautions for the children at risk. A number of data preprocessing processes were applied to increase the accuracy rate of the prediction model. A wrapper method for feature subset selection was applied to find the optimal subset of features. After that, three popular data mining algorithms (decision tree, random forest, and naive Bayes) were used and compared in terms of classification accuracy rate. In addition, this study also investigates the effects of two different grade categorizations on data mining: five-level grade categorization and binary grade categorization.

The remainder of this paper is organized as follows. In Section 2, the previous studies in this field are mentioned. In Section 3, the methods used in this study are briefly explained to provide a comprehensive understanding of the research concepts. In Section 4, experimental studies are presented with dataset description, data preprocessing, and experimental result subtitles. Finally, conclusion and the direction for future research are given in Section 5.

2. Related work

Predicting students' academic performance is one of the main topics of educational data mining [1, 2]. With the advancement of technology, technological investments in the field of education have increased. Along with technological developments, e-Learning platforms such as web-based online learning and multimedia technologies have evolved, and both learning costs have decreased, and time and space limitations have been eliminated [3]. The increase of online course trainings and the increase of online transactions and interactive transactions in schools led to the increase of digital data in this field. Costa (2017) emphasized the data about the failure rate of the students; the educators were concerned and raised important questions about the failure prediction [4].

Estimating students' performances becomes more difficult because of the large volume of data in training databases [5]. Descriptive statistical analysis can be effectively used to provide the basic descriptive information of a given set of data [6]. However, this alone is not always enough. To inform the instructors and students early, students may be able to identify early, using estimated modeling methods [7]. It is useful to classify university students according to their potential academic performance in order to increase success rates and to manage the resources well [8]. The large growth of electronic data from universities leads to an increase in the need to obtain meaningful information from these large amounts of data [9]. By using data mining techniques on education data, it is possible to improve the quality of education processes [10].

Until now, data mining algorithms have been applied on various different educational fields such as engineering education [11], physical education [12], and English language education [13]. Some studies have focused on high school students [14], while some of them have interested in higher education [15]. Whereas some data mining studies have focused on the prediction of student performance [16], some studies have investigated the instructor performance [17].

3. Method

The increase in digitalization caused us to have plenty of data in every field. Having too much data is getting worth if we know how to use it. *Data mining* aims to access knowledge from data using various machine learning techniques. With data mining, it becomes possible to establish the relationships between the data and make accurate predictions for the future. One of the application areas of data mining is education. *Data mining in education* is the field that allows us to make predictions about the future by examining the data obtained so far in the field of education by using machine learning techniques. There are basically three data mining methods: *classification*, *clustering*, and *association rule mining*. In this study, we focus on the classification task.

The methods to be used in data mining may differ depending on the field of study and the nature of the data we have. In this study, three well-known classification algorithms (decision tree, random forest, and naive Bayes) were employed on the educational datasets to predict the final grades of students.

3.1 Naive Bayes

Naive Bayes classifiers are a family of algorithms. These classifiers are based on Bayes' Theorem, which finds the possibility of a new event based on previously occurring events. Each classification is independent of one another but has a common principle.

3.2 Decision tree

A decision tree uses a tree like graph. Decision trees are like flowchart but not noncyclic. The tree consists of nodes and branches. Nodes and branches are arranged in a row. Root node is on the top of a tree and represents the entire dataset. Entropy is calculated when determining nodes in a tree. It models decisions with efficacy, results, and resource costs. In this study, decision tree technique is preferred because it is easy to understand and interpret.

3.3 Random forest

Random forest is an ensemble learning algorithm. It is a supervised classification method. It consists of randomly generated many decision trees. The established forest is formed by the decision trees community trained by the bagging method, which is one of the ensemble methods. Random forest creates multiple decision trees and combines them to achieve a more accuracy rates and stable prediction.

Figure 1 illustrates the workflow of data mining model for classification. In the first step, feature selection algorithms are applied on the educational data. Next, classification algorithms are used to build a good model which can accurately map inputs to desired outputs. The model evaluation phase provides feedback to the feature selection and learning phases for adjustment to improve classification performance. Once a model is built, then, in the second phase, it is used to predict label of new student data.

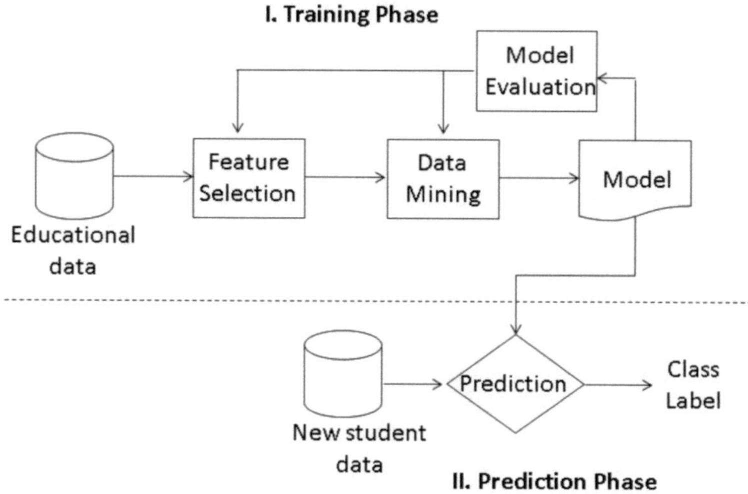

Figure 1.
Flowchart of the data mining model.

4. Experimental studies

In this study, the feature subset selection and classification operations were conducted by using WEKA open-source data mining software [18]. In each experiment, 10-fold cross-validation was performed to evaluate the classification models. The classification accuracy of the algorithm for the test dataset was measured as given in Eq. 1:

$$\text{accuracy}(T) = \frac{\sum_{i=1}^{|T|} \text{eval}(t_i)}{|T|} \qquad \text{eval}(t) = \begin{cases} 1, \text{if classify}(t) = c \\ 0, \text{otherwise} \end{cases} \tag{1}$$

where T is a test set that consists of a set of data items to be classified; c is the actual class of the item t, where $t \in T$; and *classify(t)* returns the classification output of t by the algorithm.

4.1 Dataset description

In this study, two publically available datasets [19] were used to predict student performances. Both datasets were collected from secondary education of two Portuguese schools. Dataset attributes are about student grades and social, demographic, and school-related features. All data were obtained from school reports and questionnaires. The first dataset has information regarding the performances of students in Mathematics lesson, and the other one has student data taken from Portuguese language lesson. Both datasets have 33 attributes as shown in **Table 1**.

4.2 Data preprocessing

In the raw dataset, the final grade is in the range of 0–20 as with many European countries, where 0 is the worst grade and 20 is the best score. Since the final grade of the students is in the form of integer, the predicted class should be in the form of categorical values, the data needed to be transformed to categories according to a grading policy. In the study, we used and compared two different grading systems: five-level grading and binary grading systems.

We first categorized the final grade in five groups. These ranges are defined based on the Erasmus system. As shown in **Table 2**, the range 0–9 refers to grade F, which is the worst grade and corresponds to "fail" label. The others (10–11, 12–13, 14–15, and 16–20) correspond to D (sufficient), C (satisfactory), B (good), and A (excellent/very good) class labels, respectively.

To compare the results, we also categorized the final grade as "passed" and "fail." As shown in **Table 3**, the range of 0–9 corresponds to F, and it means "fail"; the range of 10–20 refers to A, B, C, and D, and it means "pass."

4.3 Experimental results

As a preprocessing operation, the final grade attribute was categorized according to two different grading systems, before classification. As a result, we have created two versions of both datasets. Both math and Portuguese datasets were available in both five-level and binary grading versions. Hence, we have the chance to compare the results of these versions.

In the first experiment, three algorithms [decision tree (J48), random forest, and naive Bayes] were compared on the five-level grading version and binary version of

Feature	Description	Type	Values
Sex	The gender of the student	Binary	Female or male
Age	The age of the student	Numeric	From 15 to 22
School	The school of the student	Binary	GP (Gabriel Pereira) or MS (Mousinho da Silveira)
Address	Home address type of student	Binary	Urban or rural
Pstatus	Cohabitation status of student's parent	Binary	Living together or apart
Medu	Education of student's mother	Numeric	From 0 to 4
Mjob	Job of student's mother	Nominal	Teacher, health, services, at home, others
Fedu	Education of student's father	Numeric	From 0 to 4
Fjob	Job of student's father	Nominal	Teacher, health, services, at home, others
Guardian	Guardian of student	Nominal	Mother, father, or otherd
Famsize	Size of family	Binary	"LE3" (less or equal to 3) or "GT3" (greater than 3)
Famrel	Quality of family relationships	Numeric	From 1 very bad to 5 excellent
Reason	Reason of choosing this school	Nominal	Close to home, school reputation, course preference, or others
Travel time	Travel time of home to school	Numeric	1–<15 min., 2–15 to 30 min., 3–30 min. to 1 hour, or 4–>1 hour
Study time	Study time of a week	Numeric	–< 2 hours, 2–2 to 5 hours, 3–5 to 10 hours or 4– > 10 hours
Failures	Number of past class failures	Numeric	n if 1 < =n < 3, else 4
Schoolsup	Extra educational school support	Binary	Yes or no
Famsup	Family educational support	Binary	Yes or no
Activities	Extracurricular activities	Binary	Yes or no
Paid class	Extra paid classes	Binary	Yes or no
Internet	Internet access at home	Binary	Yes or no
Nursery	Attended nursery school	Binary	Yes or no
Higher	Wants to take higher education	Binary	Yes or no
Romantic	With a romantic relationship	Binary	Yes or no
Free time	Free time after school	Numeric	From 1 (very low) to 5 (very high)
Go out	Going out with friends	Numeric	From 1 (very low) to 5 (very high)
Walc	Alcohol consumption of weekend	Numeric	From 1 (very low) to 5 (very high)
Dalc	Alcohol consumption of workday	Numeric	From 1 (very low) to 5 (very high)
Health	Status of current health	Numeric	From 1 (very low) to 5 (very high)
Absences	Number of school absences	Numeric	From 0 to 93
G1	Grade of first period	Numeric	From 0 to 20
G2	Grade of second period	Numeric	From 0 to 20
G3	Grade of final period	Numeric	From 0 to 20

Table 1.
The main characteristics of the dataset.

1	2	3	4	5
Excellent/very good	Good	Satisfactory	Sufficient	Fail
16–20	14–15	12–13	10–11	0–9
A	B	C	D	F

Table 2.
Five-level grading categories.

Pass	Fail
10–20	0–9
A, B, C, D	F

Table 3.
Binary fail/pass category.

the Portuguese dataset. As shown in **Table 4**, the best performance for the five-level grading version for this dataset was obtained with an accuracy rate of 73.50% with the random forest algorithm. However, this accuracy rate was increased with binary grading version of this dataset. In the dataset, where the final grade is categorized in binary form (passing or failing), the accuracy rate was increased to 93.07%.

The performances of three classification algorithms on mathematics datasets (five-level and binary label dataset versions) are shown in **Table 5**. The best results for five-level grading version were obtained with the decision tree (J48) algorithm with an accuracy rate of 73.42%. The best accuracy rate 91.39% for binary dataset version was obtained with the random forest ensemble method.

As a second experiment, we made all comparisons after dataset preprocessing, in other terms, after feature subset selection. Hence, the most appropriate attributes were selected by using wrapper subset method to increase the accuracy rates.

One of the important steps to create a good model is attribute selection. This operation can be done in two ways: first, select relevant attributes, and second, remove redundant or irrelevant attributes. Attribute selection is made to create a

Algorithm	Five-level grading	Binary grading (P/F)
Decision tree (J48)	67.80%	91.37%
Random forest	**73.50%**	**93.07%**
Naive Bayes	68.26%	88.44%

(accuracy values, bold – best model).

Table 4.
Classification accuracy rates for the Portuguese lesson dataset.

Mathematics	Five-level grading	Binary grading (P/F)
Decision tree (J48)	**73.42%**	89.11%
Random forest	71.14%	**91.39%**
Naive Bayes	70.38%	86.33%

(accuracy values, bold – best model).

Table 5.
Classification accuracy rates for the mathematics lesson dataset.

simple model, to create a model that is easier to interpret, and to find out which features are more important for the results. Attribute selection can be done using filters and wrapper methods. In this study, we use the wrapper method, because it generally produces better results. This method has a recursive structure. The process starts with selecting a subset and induces the algorithm on that subset. Then evaluation is made according to the success of the model. There are two options in this assessment. The first option returns to the top to select a new subset, the second option uses the currently selected subset.

In **Table 6**, the accuracy rates were compared before and after the attribute selection process for the Portuguese dataset for five-level grade version. Thanks to the wrapper subset method, the accuracy rate of the J48 algorithm has increased from 67.80 to 74.88% with the selected attributes. This accuracy rate increased from 68.26 to 72.57% for naive Bayes algorithm. For the random forest method where we get the best accuracy results, the accuracy rate has increased from 73.50 to 77.20%.

In **Table 7**, the accuracy rates were compared before and after the attribute selection process for the mathematics dataset for five-level grading version. In this dataset, attribute selection significantly increased our accuracy. Here, unlike Portuguese language dataset, the best jump was obtained with J48 algorithm and search forward technique in wrapper method. In this way, the accuracy rate increased from 73.42 to 79.49%. A close result was obtained with the search backward technique and accuracy increased from 73.42 to 78.23%. Through this way, naive Bayes and random forest methods also increased significantly. This method increased the

Feature selection	Wrapper subset (J48)	Wrapper subset (naive Bayes)	Wrapper subset (random forest)
Before	67.80%	68.26%	73.50%
After	74.88%	72.57%	**77.20%**
Selected features	Search backward: age, famsize, Mjob, schoolsup, paid, internet, go out, health, G1, G2	Search backward: travel time, romantic, free time, health, G1, G2	School, Travel time, G2

The obtained classification accuracy rates for the Portuguese lesson dataset with five-level grading system. (accuracy values, bold – best model).

Table 6.
Before and after feature selection with five-level grading system

Feature selection	Wrapper subset (J48)	Wrapper subset (J48)	Wrapper subset (naive Bayes)	Wrapper subset (random forest)
Before	73.42%	73.42%	70.38%	71.14%
After	78.23%	**79.49%**	74.18%	78.99%
Selectedfeatures	Search backward: age, pstatus, Medu, Fedu, Fjob, failures, schoolsup, paid, activities, famrel, Dalc, Walc, G2	Search forward: sex, Mjob, Walc, G2	Famsize, Medu, Fjob, activities, higher, romantic, free time, G2	Famsize, Fedu, schoolsup, paid, activities, higher, romantic, Walc, absences, G1, G2

The obtained classification accuracy rates for the mathematics lesson dataset with five-level grading system. (accuracy values, bold – best model).

Table 7.
Before and after feature selection with binary grading system.

Feature selection	Wrapper subset (J48)	Wrapper subset (naive Bayes)	Wrapper subset (random forest)
Before	91.37%	88.44%	93.07%
After	91.99%	89.68%	**93.22%**
Selected Features	School, age, address, Medu, Fjob, travel time, study time, schoolsup, nursery, higher, famrel, free time, G1, G2	Sex, age, Pstatus, Fedu, Mjob, Fjob, reason, failures, famsup, paid, higher, Internet, romantic, go out, health, absences, G1, G2	School, sex, age, address, famsize, Pstatus, Medu, Mjob, Fjob, reason, guardian, travel time, study time, failures, schoolsup, famsup, paid, activities, higher, Internet, romantic, famrel, free time, go out, Dalc, Walc, health, absences, G1, G2

The obtained classification accuracy rates for the Portuguese lesson dataset with binary grading system. (accuracy values, bold – best model).

Table 8.
Before and after feature selection.

Feature selection	Wrapper subset (J48)	Wrapper subset (naive Bayes)	Wrapper subset (random forest)
Before	89.11%	86.33%	91.39%
After	90.89%	89.11%	**93.67%**
Selected features	School, age, address, Medu, Fedu, guardian, failures, schoolsup, famsup, Internet, romantic, famrel, free time, G1, G2	Sex, age, Pstatus, Fedu, Mjob, Fjob, reason, failures, famsup, paid, higher, Internet, romantic, go out, health, absences, G1, G2	Address, famsize, Fedu, Mjob, Fjob, reason, guardian, study time, schoolsup, higher, famrel, go out, absences, G2

The obtained classification accuracy rates for the mathematics lesson dataset with binary grading system. (accuracy values, bold – best model).

Table 9.
Before and after feature selection.

accuracy rate of naive Bayes method from 70.38 to 74.18%. Random forest result is increased from 71.14 to 78.99%. These results show that attribute selection with this wrapper subset method also works in this dataset.

In **Table 8**, the results of the wrapper attribute selection method before and after the application to the Portuguese binary version are compared. There was no significant increase in accuracy. The best results were obtained with random forest. The best jump was experienced by the naive Bayes method but did not reach the random forest value. Naive Bayes result has risen from 88.44 to 89.68%. Random forest maintained the high accuracy achieved before the attribute selection and increased from 93.07 to 93.22%.

After successful results in five-level grade versions, we tried the same attribute selection method in binary label version dataset. **Table 9** shows the accuracy values before and after the wrapper attribute selection for the mathematical binary dataset version. Because the accuracy of the binary version is already high, the jump is less than the five-level grades. But again, there is a nice increase in accuracy. The accuracy rate of the J48 algorithm was increased from 89.11 to 90.89%, while the naive Bayes result was increased from 86.33 to 89.11%. As with the mathematics five-level grade dataset, the best results were obtained with random forest in binary label dataset. Accuracy rate increased from 91.39 to 93.67%.

As a result, it can be possible to say that accuracy rates have changed positively in all trials using wrapper subset attribute selection method.

5. Conclusion and future work

This paper proposes the application of data mining techniques to predict the final grades of students based on their historical data. Three well-known classification techniques (decision tree, random forest, and naive Bayes) were compared in terms of accuracy rates. Wrapper feature subset selection method was used to improve the classification performance. Preprocessing operations on the dataset, categorizing the final grade field into five and two groups, increased the percentage of accurate estimates in the classification. The wrapper attribute selection method in all algorithms has led to a noticeable increase in accuracy rate. Overall, better accuracy rates were achieved with the binary class method for both mathematics and Portuguese dataset.

In the future, different feature selection methods can be used. In addition, different classification algorithms can also be utilized on the datasets.

Author details

Ferda Ünal
The Graduate School of Natural and Applied Sciences, Dokuz Eylul University, Izmir, Turkey

*Address all correspondence to: ferda.balci@ceng.deu.edu.tr

References

[1] Fan Y, Liu Y, Chen H, Ma J. Data mining-based design and implementation of college physical education performance management and analysis system. International Journal of Emerging Technologies in Learning. 2019;**14**(06):87-97

[2] Guruler H, Istanbullu A. Modeling student performance in higher education using data mining. Studies in Computational Intelligence. 2014;**524**: 105-124

[3] Hu YH, Lo CL, Shih SP. Developing early warning systems to predict students' online learning performance. Computers in Human Behavior. 2014; **36**:469-478

[4] Costa EB, Fonseca B, Santana MA, de Araújo FF, Rego J. Evaluating the effectiveness of educational data mining techniques for early prediction of students' academic failure in introductory programming courses. Computers in Human Behavior. 2017;**73**: 247-256

[5] Shahiri AM, Husain W. A review on predicting student's performance using data mining techniques. Procedia Computer Science. 2015;**72**:414-422

[6] Fernandes E, Holanda M, Victorino M, Borges V, Carvalho R, Van Erven G. Educational data mining: Predictive analysis of academic performance of public school students in the capital of Brazil. Journal of Business Research. 2019;**94**:335-343

[7] Marbouti F, Diefes-Dux HA, Madhavan K. Models for early prediction of at-risk students in a course using standards-based grading. Computers in Education. 2016;**103**:1-15

[8] Miguéis VL, Freitas A, Garcia PJ, Silva A. Early segmentation of students according to their academic performance: A predictive modelling approach. Decision Support Systems. 2018;**115**:36-51

[9] Asif R, Merceron A, Ali SA, Haider NG. Analyzing undergraduate students' performance using educational data mining. Computers in Education. 2017;**113**:177-194

[10] Rodrigues MW, Isotani S, Zárate LE. Educational Data Mining: A review of evaluation process in the e-learning. Telematics and Informatics. 2018;**35**(6): 1701-1717

[11] Buenano-Fernandez D, Villegas-CH W, Lujan-Mora S. The use of tools of data mining to decision making in engineering education—A systematic mapping study. Computer Applications in Engineering Education. 2019;**27**(3): 744-758

[12] Zhu S. Research on data mining of education technical ability training for physical education students based on Apriori algorithm. Cluster Computing. 2019;**22**(6):14811-14818

[13] Lu M. Predicting college students English performance using education data mining. Journal of Computational and Theoretical Nanoscience. 2017; **14**(1):225-229

[14] Marquez-Vera C, Cano A, Romero C, Noaman AYM, Mousa FH, Ventura S. Early dropout prediction using data mining: A case study with high school students. Expert Systems. 2016;**33**(1):107-124

[15] Amjad Abu S, Al-Emran M, Shaalan K. Factors affecting students' performance in higher education: A systematic review of predictive data mining techniques. Technology, Knowledge and Learning. 2019;**24**(4): 567-598

[16] Fujita H. Neural-fuzzy with representative sets for prediction of student performance. Applied Intelligence. 2019;**49**(1):172-187

[17] Agaoglu M. Predicting instructor performance using data mining techniques in higher education. IEEE Access. 2016;**4**:2379-2387

[18] Hall M, Frank E, Holmes G, Pfahringer B, Reutemann P, Witten IH. The WEKA data mining software: An update. ACM SIGKDD explorations newsletter. 2009

[19] Cortez P, Silva A. Using data mining to predict secondary school student performance. In: Brito A, Teixeira J, editors. Proceedings of 5th Annual Future Business Technology Conference. tpPorto: EUROSIS-ETI; 2018. pp. 5-12

Chapter 10

Tracer Transport in a Homogeneous Porous Medium: Experimental Study and Acquisition Data with LabVIEW

Sana Dardouri and Jalila Sghaier

Abstract

This work represent the incorporation of information procurement (DAQ) equipment and programming to acquire information (LabVIEW) as well as real-time transport to show parameter appraises with regard to subsurface stream and transport issues. The main objective is to understand the mechanism of water and solute transfer in a sandy medium and to study the effect of some parameters on the transport of an inert tracer. In order to achieve this objective, a series of experiments were carried out on a soil column equipped with a tensiometer to monitor the state of saturation of the medium and by two four-electrode probes for measuring the electrical conductivity in the porous medium.

Keywords: tracer test experiments, groundwater contaminant, transport in porous media

1. Introduction

The comprehension of contaminant destiny in groundwater conditions is of high enthusiasm for supply and the executives of water assets in urban regions. Contaminant discharges invade through the dirt to cross the vadose zone and achieve the water table where they spread after the specific stream bearings and hydrodynamic states of groundwater bodies. Localization and checking of contaminants is the primary essential advance to remediation systems [1, 2].

In any case, exact data are generally compelled by the absence of thickness of inspecting areas which are illustrative of the region of the boreholes yet do not render of nearby heterogeneities and particular stream headings of the crest [3].

A slug of solutes (tracers) promptly infused into permeable media with a uniform stream field is normally alluded to as the slug-tracer test. The injected tracer will go through permeable media as a pulse with a peak concentration eventually after injection. This sort of test is utilized generally to determinate contaminant transport parameters in permeable media or subsurface conditions [4, 5]. The transport parameters including porosities, pore velocities, and dispersivities are imperative to examine the fate and transport of the contaminants and colloid in permeable media and groundwater [6–9].

Many studies showed that the type of array and the sequence of measurements incredibly affected the shape and intensity of the resistivity contrasts ascribed to tracer temporal spreads [10, 11].

2. Experimental study

2.1 Materials and methods

The experimental setup (**Figure 1**) consists of a cylindrical glass column of 36 cm long and 7.5 cm in diameter. The column is closed below with a plastic cover having a hole at the center of diameter 1 cm and provided with a filter grid preventing the passage of the solid phase beyond the column. The pressure within the column is measured using a tensiometer located 5 cm from the bottom of the column. This blood pressure monitor is of type Soilmoisture model 2100F. There are also two four-electrode probes located 5 and 30 cm from the base of the column. These probes make it possible to follow the transport of a tracer by measuring the electrical conductivity in the ground.

2.1.1 Pressure measurement

The 2100F Soilmoisture Tester (**Figure 2**) is an instrument designed to measure soil pressure potentials. This model is ideal for laboratory measurements such as measurement of soil suction at fixed depth in a soil column. This blood pressure monitor consists mainly of a porous ceramic stem (ceramic is rigid and permeable), a ventilation tube, a plastic body, and a Bourdon manometer. This circulatory strain screen comprises for the most part of a permeable earthenware stem (artistic is unbending and penetrable), a ventilation tube, a plastic body, and a Bourdon manometer.

The operating principle of this blood pressure monitor is simple. Indeed, when the porous ceramic saturated with water is placed in the unsaturated soil, a water potential gradient appears between the interior of the porous ceramic and the soil. These results in a transfer of water to the soil which exerts a depression (suction)

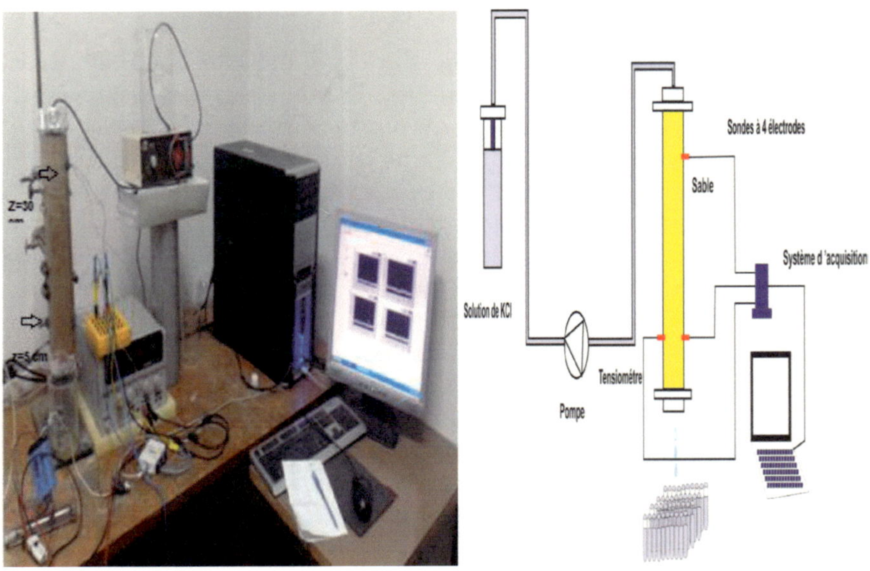

Figure 1.
Experimental device.

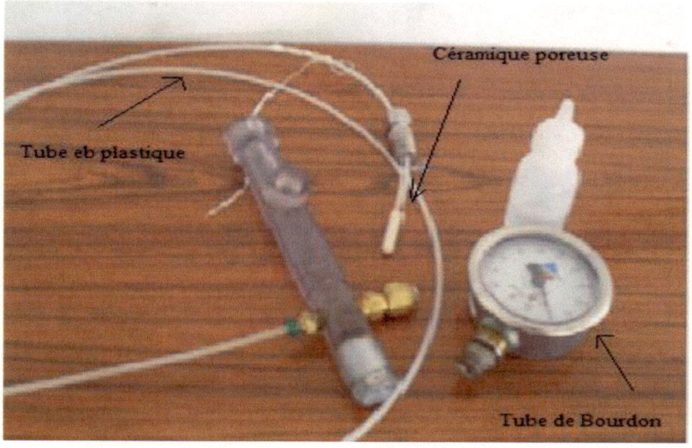

Figure 2.
"Soilmoisture 2001F" blood pressure monitor.

on the water contained in the tensiometer. The transfer of water takes place through the porous wall of the ceramic and can only exist if the liquid phase is continuous from the ground, the wall of the rod, and the inside of the tube. It is necessary to calibrate the monitor before use to ensure proper function. This calibration is performed as follows:

The first step is to immerse the porous ceramic in the water. At the same time, remove the drain screw and fill the plastic tube. It should be noted that filling is done slowly to prevent the occlusion of a large unwanted air volume in the nylon tube. The tube is continued to fill until a flow of water without air bubbles at the aeration tube is obtained. To purge all the air bubbles and make sure that the tube is completely filled with water, the drain screw is tightened, and the moisture present in the porous ceramic is removed.

As the water evaporates on the surface of the porous ceramic, an increase in the Bourdon tube needle is observed due to the increase in vacuum in the tube. After an hour or two, the manometer reading increases to a value equal to or greater than 60 centibar. This is an accumulation of air volume trapped in the nylon tube and the plastic tube. To remove this accumulated air, first tap the plastic tube to remove air bubbles as much as possible, then remove the lid of the plastic tube, and add water as previously done. The inner nylon tube is then trimmed so that it does not exceed 6.35 mm. After balancing, tighten the drain screw and cover.

Acquisition of pressure through a current transducer (**Figure 3**) is accomplished through the NI-DAQ 6009 acquisition card (**Figure 4**). The transducer output

Figure 3.
Current transducer.

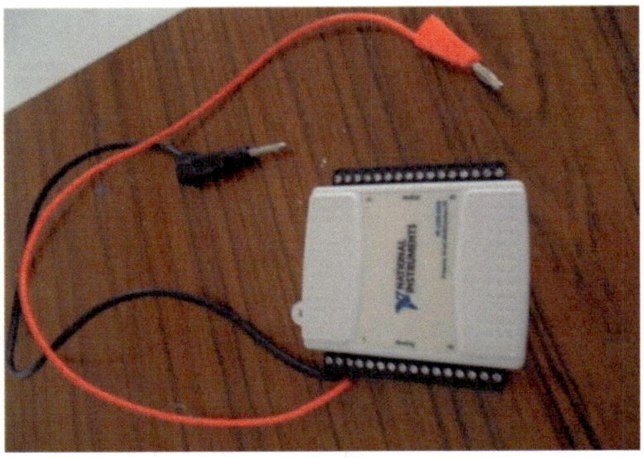

Figure 4.
NI 6009 acquisition board.

is connected to the analog input of the board. The LabVIEW software supplied with the acquisition card allows, thanks to its graphic environment, to obtain the desired measurements. Another solution for the acquisition is the direct use of the NI-DAQmx acquisition card driver with a small code on MATLAB R2012, certainly less developed but which also allows to acquire the pressure data.

Using NI-DAQmx, one must first choose the type of the property to be measured (voltage, current, strain gauge/pressure, temperature, etc.) and then start the acquisition by choosing the number of samples to be measured and the sampling period.

It should be noted that the current transducer used is of the 4–20 mA current loop type. That is, this device measures the pressure by converting it into current such that the minimum value (0cbar) corresponds to 4 mA and the maximum value (100 cbar) corresponds to 20 mA. Measurements can be voltage measurements by connecting the current transducer to a 500 Ω resistor.

2.1.2 Concentration measurement

Two four-electrode probes for the measurement of the electrical conductivity in the soil were carried out within the ENIM, in collaboration with the electrical engineering department. For each probe (**Figure 5**), a printed circuit has been designed which has four equidistant copper surfaces of 6 mm representing the four electrodes. The purpose of manufacturing these probes is to measure the electrical conductivity in the soil as a function of time at a given depth.

The operating principle of these probes consists of sending an alternating electric current into the ground through the two surface electrodes and measuring the potential difference by means of two inner electrodes.

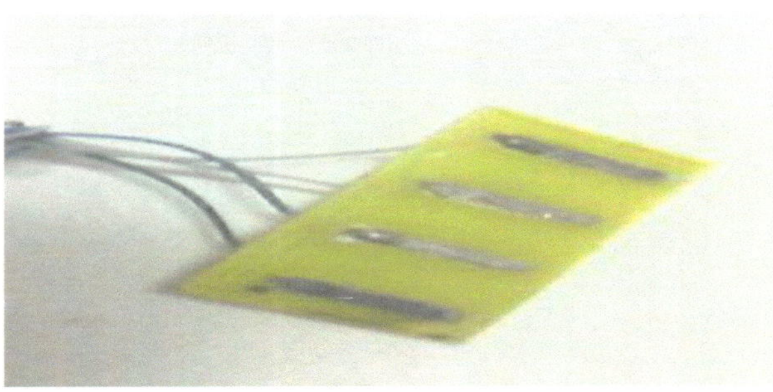

Figure 5.
Four-electrode probe for measuring electrical conductivity.

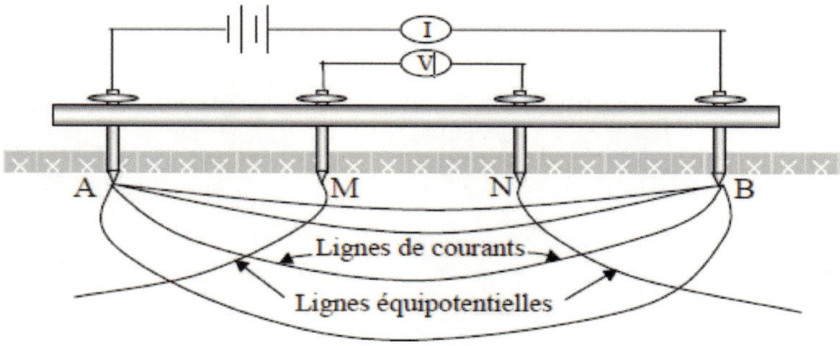

Figure 6.
Schematic diagram of the Wenner model.

The method of four electrodes was chosen from several generally non-destructive electrical methods because its principle is simple and the application disrupts very little the flow in the ground. The geometry of these probes is based on the Wenner configuration (**Figure 6**).

The electrical resistivity of the ground is written as

$$\rho = \frac{\Delta V}{I} . 2\pi a \tag{1}$$

where a is the distance that separates each electrode from the other.

The apparent electrical conductivity is then determined by

$$\sigma_a = \frac{kf}{R_s} \tag{2}$$

where k = a*(1/4) is a constant; f is the temperature correction factor; and Rs is the electrical resistance equal to ΔV/I.

It is essential to know the electrical behavior of the manufactured probe so that we can draw good results and especially choose the appropriate acquisition system. Thus, several tests were carried out with sand and NaCl solution. The probe is pushed into the sand and the solution is injected. At the same time, the response of the probe is visualized on an oscilloscope. It has been concluded that the probe behaves similar to a capacitive impedance since the current and voltage output signals are out of phase (**Figure 7**). In addition, the frequency behavior performed with a GBF also confirms this.

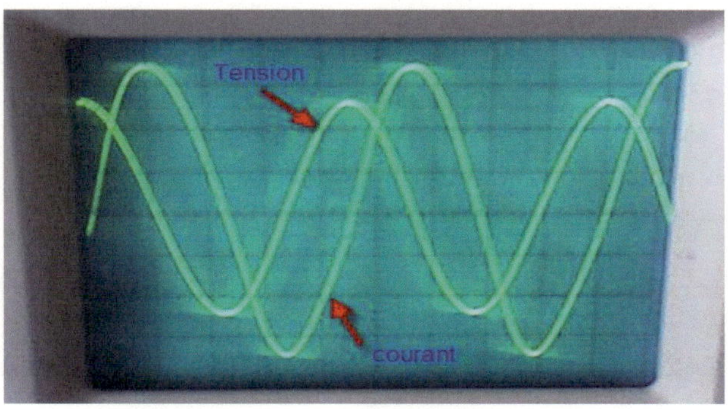

Figure 7.
Electrical behavior of the probe.

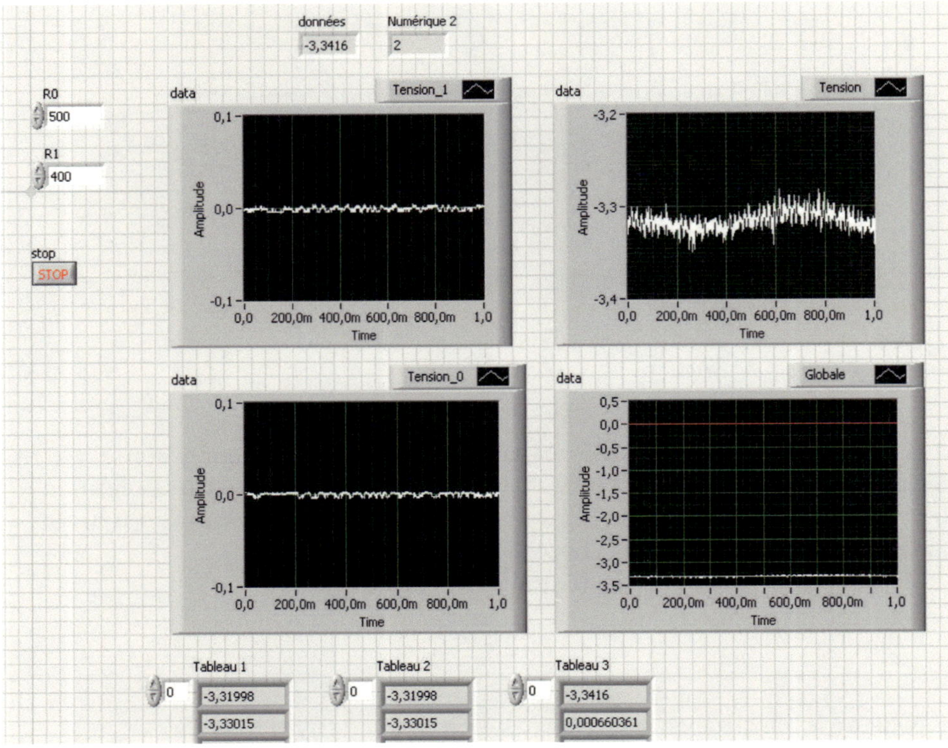

Figure 8.
Front side of the program under LabVIEW.

2.2 Data acquisitions

Regarding the acquisition of the data signals, the LabVIEW software was manipulated to record the results acquired by the three current sensors (pressure sensor and two electrical conductivity sensors). We chose an average acquisition frequency corresponding to 1800 points for the tracer injection phase and 720 points for the leaching phase. The acquisition principle diagram under LabVIEW (program front panel (**Figure 8**)) is as follows:

2.2.1 Synthesis of the experiments carried out on the soil column

A total of four experiments was performed on homogeneous porous media under saturated conditions with a slot injection of a tracer. The tests carried out are summarized in **Table 1**. The main tasks performed are mentioned and dissolved for each experiment.

- The column is filled with initially dry soil, and the sand is well sanded to prevent maximum entrapment of air bubbles in the porous medium.
- During filling, two four-electrode probes are introduced at two levels of the column, z = 5 and z = 30 cm, to measure the electrical conductivity in the soil. The tensiometer is also introduced at a height of 5 cm from the bottom of the column.
- A well-defined volume of water is fed at the top of the column for saturation of the medium.

	Experiences			
	A	**B**	**C**	**D**
Type of injection	Slot	Slot	Slot	Slot
Average injection rate ($l.min^{-1}$)	0.044	0.053	0.053	0.07
Tracer used	NaCl	KCl	KCl	KCl
Tracer concentration ($g.l^{-1}$)	2.8	0.74	0.74	0.74
Medium studied	Sousse sand	Monastir sand	Sousse sand	Sousse sand
Water status of the saturated medium	Distilled water	Distilled water	Distilled water	Distilled water
Medium saturation condition	Saturated	Saturated	Saturated	Saturated

Table 1.
Summary of all experiments carried out in the soil column.

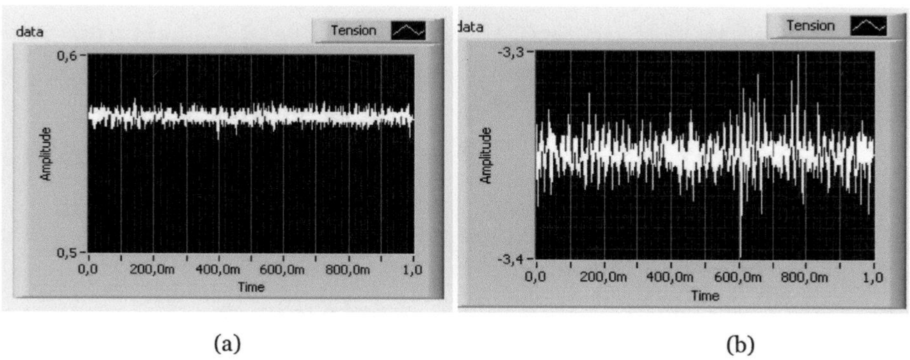

Figure 9.
Recording tensions acquired by the tensiometer for (a) a saturated medium and (b) an unsaturated medium.

- The pressure load within the column is monitored until it reaches a positive and stable value indicating saturation of the medium (**Figure 9a**).
- Once assured that our medium is saturated, we begin the tracer injection with a peristaltic pump (Roth Cyclo I) at a constant average flow rate.

3. Results and discussions

3.1 Evolution of electrical conductivity

The four experiments are described well, through the evolution of the electrical conductivity in the soil and the phenomena of convection and dispersion. Indeed, moving away from the upper base of the column (tracer injection point), that is, to say down the height of the column, the electrical conductivity decreases. It is clear, by comparing two curves of the same sample at z = 5 cm and z = 30 cm, that there is a significant delay (**Figure 10**).

This delay and this small peak are the reasons for the quasi-total dispersion of the tracer in the soil water and, subsequently, the decrease of the electrical conductivity. The high value of the electrical conductivity in the Sousse sand in experiment A (**Figure 10a**) is justified by the high concentration of the NaCl tracer injected into the column.

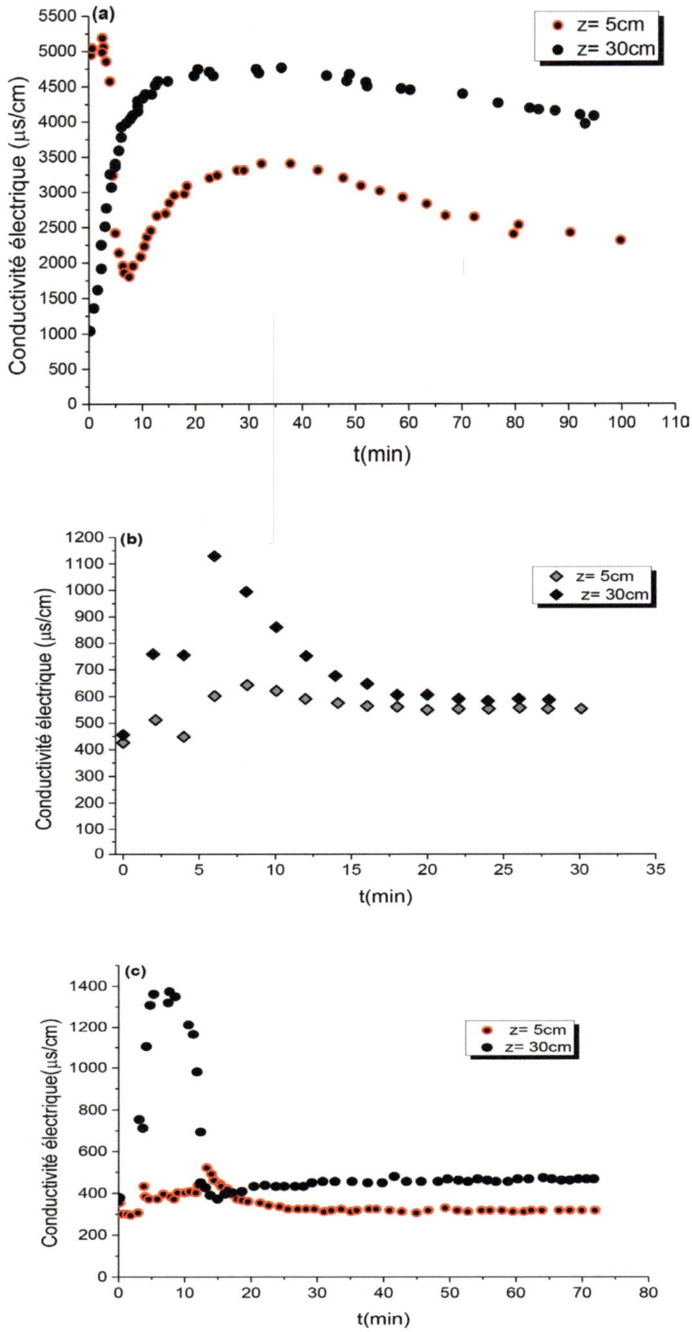

Figure 10.
Evolution of electrical conductivity: (a) experiment A, (b) experiment B, and (c) experiment C.

3.2 Effect of the hydraulic conductivity of the medium

It can be seen from **Figure 11** that the electrical conductivity curve for slot tracer injection in the Sousse sand has the same behavior as the Monastir sand. The peaks of the ascending part of two curves are at very close moments. This indicates that the hydrodynamic characteristics of two media are close and that there is no interaction between the tracer and the medium [12]. The peak shape of the curve of the Monastir sand sample proves that during the flow, there is no preferential

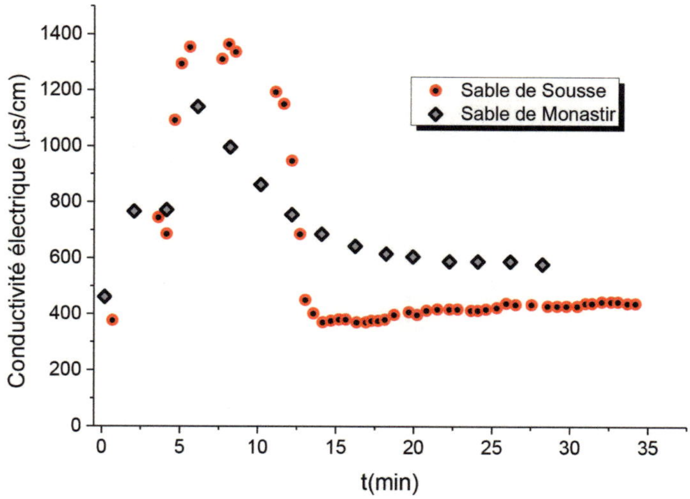

Figure 11.
Effect of medium permeability: comparison between experiments B and C.

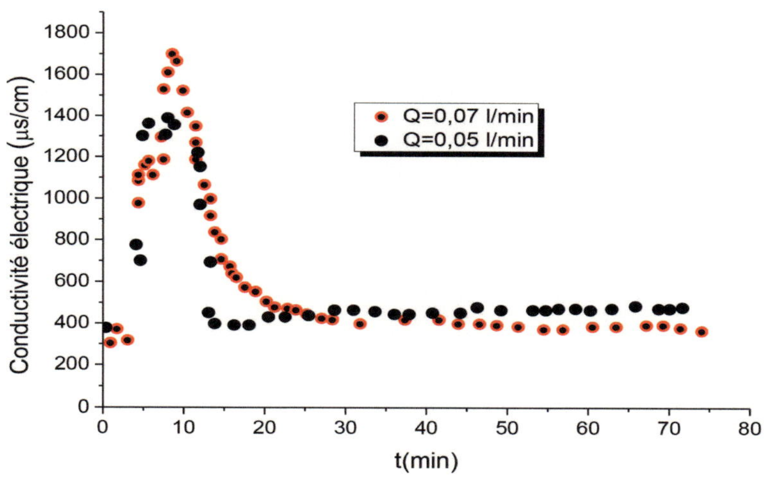

Figure 12.
Effect of average tracer injection rate: comparison between experiments C and D.

path creation. The small offset in ordinates for the two curves can be justified by the nature of sandy environment. The sand of Sousse is thinner and has a lower saturation hydraulic conductivity than the sand of Monastir. The low hydraulic conductivity of the Sousse sand results in greater retention of the tracer and an increase in the measured electrical conductivity.

3.3 Injection flow effect

The influence of injection rate of a tracer KCl on the curves of the electrical conductivity in a sandy medium (sand of Sousse region) at a height z = 30 cm from the bottom of the column is studied. The two selected flow rates are high and correspond to flow rates in the column of 0.327 and 0.458 cm/s. By increasing the flow, the tracer appears faster and disappears more slowly. Indeed, the higher the flow rate, the greater the dispersion phase of the breakthrough curve [13]. In our case, the two flow values are close, which explains why the variation has no significant effect on the shape of the tracer restitution curve (**Figure 12**).

4. Conclusion

In this chapter, we studied the transport of two inert tracers in a homogeneous porous medium (sand). The effects of injection rate and permeability of the medium on the evolution of the tracer elution curve were examined.

Author details

Sana Dardouri* and Jalila Sghaier
National Engineering School of Monastir, Monastir, Tunisia

*Address all correspondence to: sanadardouri_en@yahoo.fr

References

[1] Herrera GS, Pinder GF. Space-time optimization of groundwater quality sampling networks. Water Resources Research. 2005;**41**:W12407

[2] Meyer PD, Valocchi AJ, Eheart JW. Monitoring network design to provide initial detection of groundwater contamination. Water Resources Research. 1994;**30**(9):2647-2659

[3] Kim KH, Lee KK. Optimization of groundwater-monitoring networks for identification of the distribution of a contaminant plume. Stochastic Environmental Research and Risk Assessment. 2007;**21**(6):785-794

[4] Bear J. Some experiments in dispersion. Journal of Geophysical Research. 1961;**66**(8):2455-2467

[5] Mackay DM, Freyberg DL, Roberts PV. A natural gradient experiment on solute transport in a sand aquifer. 1. Approach and overview of plume movement. Water Resources Research. 1986;**22**(13):2017-2029

[6] Boy-Roura M et al. Towards the understanding of antibiotic occurrence and transport in groundwater: Findings from the Baix Fluvia alluvial aquifer (NE Catalonia, Spain). Science of the Total Environment. 2018;**612**:1387-1406

[7] Han B, Liu W, Zhao X, Cai Z, Zhao D. Transport of multi-walled carbon nanotubes stabilized by carboxymethyl cellulose and starch in saturated porousmedia: Influences of electrolyte, clay and humic acid. Science of the Total Environment. 2017;**599**:188-197

[8] Liang X, Zhan H, Liu J, Dong G, Zhang YK. A simple method of transport parameter estimation for slug injecting tracer tests in porous media. Science of the Total Environment. 2018;**644**:1536-1546

[9] Ma J et al. Enhanced transport of ferrihydrite colloid by chain-shaped humic acid colloid in saturated porous media. Science of the Total Environment. 2018;**621**:1581-1590

[10] Bellmunt F, Marcuello A, Ledo J, Queralt P. Capability of cross-hole electrical configurations for monitoring rapid plume migration experiments. Journal of Applied Geophysics. 2016;**124**:73-82

[11] Lekmine G, Auradou H, Pessel M, Rayner JL. Quantification of tracer plume transport parameters in 2D saturated porous media by cross-borehole ERT imaging. Journal of Applied Geophysics. 2017;**139**:291-305

[12] Bayard R. Etude de l'adsorption/désorption de polluants organiques dans les sols: Approche méthodologique et application au pentachlorophénol et aux hydrocarbures aromatiques polycycliques. Diss. Lyon, INSA; 1997

[13] Dalla Costa C. Transferts de traceur en milieu poreux consolidé et milieu poreux fissuré: Expérimentations et Modélisations [Doctoral dissertation]. Université Joseph-Fourier-Grenoble I; 2007

Chapter 11

Data Mining and Fuzzy Data Mining Using MapReduce Algorithms

Poli Venkata Subba Reddy

Abstract

Data mining is knowledge discovery process. It has to deal with exact information and inexact information. Statistical methods deal with inexact information but it is based on likelihood. Zadeh fuzzy logic deals with inexact information but it is based on belief and it is simple to use. Fuzzy logic is used to deal with inexact information. Data mining consist methods and classifications. These methods and classifications are discussed for both exact and inexact information. Retrieval of information is important in data mining. The time and space complexity is high in big data. These are to be reduced. The time complexity is reduced through the consecutive retrieval (C-R) property and space complexity is reduced with blackboard systems. Data mining for web data based is discussed. In web data mining, the original data have to be disclosed. Fuzzy web data mining is discussed for security of data. Fuzzy web programming is discussed. Data mining, fuzzy data mining, and web data mining are discussed through MapReduce algorithms.

Keywords: data mining, fuzzy logic, fuzzy data mining, web data mining, fuzzy MapReduce algorithms

1. Introduction

Data mining is an emerging area for knowledge discovery to extract hidden and useful information from large amounts of data. Data mining methods like association rules, clustering, and classification use advanced algorithms such as decision tree and k-means for different purposes and goals. The research fields of data mining include machine learning, deep learning, and sentiment analysis. Information has to be retrieved within a reasonable time period for big data analysis. This may be achieved through the consecutively retrieval (C-R) of datasets for queries. The C-R property was first introduced by Ghosh [1]. After that, the C-R property was extended to statistical databases. The C-R cluster property is a presorting to store the datasets for clusters. In this chapter, C-R property is extended to cluster analysis. MapReduce algorithms are studied for cluster analysis. The time and space complexity shall be reduced through the consecutive retrieval (C-R) cluster property. Security of the data is one of the major issues for data analytics and data science when the original data is not to be disclosed.

The web programming has to handle incomplete information. Web intelligence is an emerging area and performs data mining to handle incomplete information. The incomplete information is fuzzy rather than probability. In this chapter, fuzzy web programming is discussed to deal with data mining using fuzzy logic. The fuzzy algorithmic language, called FUZZYALGOL, is discussed to design queries in data mining. Some examples are discussed for web programming with fuzzy data mining.

2. Data mining

Data mining [2–5] is basically performed for knowledge discovery process. Some of the well-known data mining methods are frequent itemset mining, association rule mining, and clustering. Data warehousing is the representation of a relational dataset in two or more dimensions. It is possible to reduce the space complexity of data mining with consecutive storage of data warehouses.

The relational dataset is a representation of data with attributes and tuples.

Definition: A relational dataset R or cluster dataset is defined as a collection of attributes $A_1, A_2, \ldots, A_m$ and tuples $t_1, t_2, \ldots, t_n$ and is represented as

$R = A_1 \text{ x } A_2 \text{ x } \ldots \text{x } A_m$

$t_i = a_{i1} \text{ x } a_{i2} \text{ x } \ldots \text{ x } a_{im}$ are tuples, where $i = 1,2,.., n$

or

$R(A_1. A_2. \ldots A_m)$. R is a relation.

$R(t_i) = (a_{i1}. a_{i2} \ldots. a_{im})$ are tuples, where $i = 1,2,.., n$

or instance, two sample datasets "price" and "sales" are given in **Tables 1** and **2**, respectively.

INo	IName	Price
I005	Shirt	100
I007	Dress	50
I004	Pants	80
I008	Jacket	60
I009	Skirt	100

Table 1.
Sample dataset "price."

INo	IName	Sales
I005	Shirt	80
I007	Dress	60
I004	Pants	100
I008	Jacket	50
I009	Skirt	80

Table 2.
Sample dataset "sales."

The lossless join of the datasets “price” and “sales” is given in **Table 3**.

INo	IName	Sales	Price
I005	Shirt	80	100
I007	Dress	60	50
I004	Pants	100	80
I008	Jacket	50	60
I009	Skirt	80	100

Table 3.
Lossless join of the price and sales datasets.

In the following, some of the methods (frequency, association rule, and clustering) are discussed.

Consider the “purchase” relational dataset given in **Table 4**.

CNo	INo	IName	Price
C001	I005	shirt	100
C001	I007	Dress	50
C003	I004	pants	80
C002	I007	dress	80
C001	I008	Jacket	60
C002	I005	shirt	100

Table 4.
Sample dataset “purchase.”

2.1 Frequency

Frequency is the repeatedly accrued data.
Consider the following query:
Find the frequently customers purchase more than one item.
SELECT P.CNo, P.INo, IName, COUNT(*)
FROM purchase P
WHERE COUNT(*)>1.
The output of this query is given in **Table 5**.

CNo	INo	COUNT
C001	I005	2
C002	I005	2

Table 5.
Frequency.

2.2 Association rule

Association rule is the relationship among the data.
Consider the following query:
Find the customers who purchase shirt and dress.
<shirt⇔ dress>
SELECT P.CNo, P.INo

FROM purchase P
WHERE IName="shirt" and IName="dress".
The output of this query is given in **Table 6**.

CNo	INo
C001	I005
C002	I005

Table 6.
Association.

2.3 Clustering

Clustering is grouping the particular data.
Consider the following query:
Group the customers who purchase dress and shirt.
The output of this query is given in **Table 7**.

CNo	INo	IName	Price
C001	I007	Dress	50
	I005	shirt	100
C002	I007	dress	80
	I005	shirt	100

Table 7.
Clustering.

3. Data mining using C-R cluster property

The C-R (consecutive retrieval) property [1, 3] is the retrieval of records of database consecutively. Suppose $R = \{r_1, r_2, \dots, r_n\}$ is the dataset of records and $C = \{C_1, C_2, \dots, C_m\}$ is the set of clusters.

The best type of file organization on a linear storage is one in which records pertaining to clusters are stored in consecutive locations without redundancy storing any data of *R*.

If there exists on such organization of *R* for *C* said to have the Consecutive Retrieval Property or C-R cluster property with respect to dataset *R*. Then C-R cluster property is applicable to linear storage.

The C-R cluster property is a binary relation between a cluster set and dataset.

R	C_1	C_2	….	C_m
r_1	1	0	…	1
r_2	0	1	;;;	0
-	-	-	…	-
-	-	-	…	-
=	-	-	…	-
r_n	1	1	…	1

Table 8.
Incidence matrix.

Suppose if a cluster in a cluster set C is relevant to the data in a dataset R, then the relevancy is denoted by 1 and the irrelevancy is denoted by 0. Thus, the relevancy between cluster set C and dataset R can be represented as $(n \times m)$ matrix, as shown in **Table 8**. The matrix is called dataset-cluster incidence matrix (CIM).

Consider the dataset for customer account given in **Table 9**.

R	CNo	IName	Sales
r_1	70001	Shirt	150
r_2	70002	Dress	30
r_3	70003	Pants	100
r_4	60001	Dress	50
r_5	60002	Jacket	75
r_6	60003	Shirt	120
r_7	60004	Dress	40

Table 9.
Storage of sales.

The dataset given in **Table 9** is reorganized in ascending order based on sorting, as shown in **Table 10**.

R	CNo	IName	Sales
r_1	70001	Shirt	150
r_6	60003	Dress	120
r_3	70003	Pants	100
r_5	60002	Dress	75
r_4	60001	Jacket	50
r_7	60004	Shirt	40
r_2	70002	Dress	30

Table 10.
Reorganizing for C-R cluster.

Consider the following clusters of queries:

C1 = Find the customers whose sales is greater than or equal to 100.

C2 = Find the customers whose sales is less than 100.

C3 = Find the customers whose sales is greater than or equal average sales.

C4 = Find the customers whose sales is less than average sales.

The CIM is given in **Table 11**.

The dataset given in **Table 11** is reorganized with sort on C_1 in descending order, as shown in **Table 12**. Thus, C_1 has C-R cluster property.

The dataset given in **Table 11** is reorganized with sort on C_2 in descending order, as shown in **Table 13**. Thus, C_2 has C-R cluster property.

The dataset given in **Table 11** is reorganized with sort on C_3 in descending order, as shown in **Table 14**. Thus, C_3 has C-R cluster property.

The dataset given in **Table 11** is reorganized with sort on C_4 in descending order, as shown in **Table 15**. Thus, C_4 has a C-R cluster property.

R	C_1	C_2	C_3	C_4
r_1	1	0	1	0
r_2	0	1	0	1
r_3	1	0	1	0
r_4	0	1	0	1
r_5	0	1	1	0
r_6	1	0	1	0
r_7	0	1	0	1

Table 11.
Cluster incidence matrix.

R	C_1
r_1	1
r_3	1
r_6	1
r_2	0
r_4	0
r_5	0
R_7	0

Table 12.
Sorting on C_1.

R	C_2
r_1	0
r_3	0
r_6	0
r_2	1
r_4	1
r_5	1
r_7	1

Table 13.
Sorting on C_2.

R	C_3
r_1	1
r_3	1
r_5	1
r_6	1
r_2	0
r_4	0
r_7	0

Table 14.
Sorting on C_3

R	C_4
r_1	0
r_3	0
r_5	0
r_6	0
r_2	1
r_4	1
r_7	1

Table 15.
Sorting on C_4.

The dataset is given for $C_1 \bowtie C_2$ has C-R cluster property (**Table 16**).

R	$C_1 \bowtie C_2$
r_1	1
r_3	1
r_6	1
r_2	1
r_4	1
r_5	1
r_7	1

Table 16.
$C_1 \bowtie C_2$.

The dataset is given for $C_3 \bowtie C_4$ has C-R cluster property (**Table 17**).

R	$C_3 \bowtie C_4$
r_1	1
r_3	1
r_5	1
r_6	1
r_2	1
r_4	1
r_7	1

Table 17.
$C_3 \bowtie C_4$.

The dataset is given for $C_1 \bowtie C_3$ has C-R cluster property (**Table 18**).
The dataset is given for $C_2 \bowtie C_4$ has C-R cluster property (**Table 19**).
The dataset is given for $C_2 \bowtie C_3$ has C-R cluster property (**Table 20**).
The cluster sets $\{C_1 \bowtie C_2, C_3 \bowtie C_4, C_1 \bowtie C_3, C_2 \text{U}\bowtie C_4, C_2 \text{U}\bowtie C_3\}$ has C-R cluster property. Thus, the cluster sets have C-R cluster properties with respect to dataset *R*.

3.1 Design of parallel C-R cluster property

The design of parallel cluster shall be studied through the C-R cluster property. It can be studied in two ways: the parallel cluster design through graph

R	$C_1 \bowtie C_3$
r_1	1
r_3	1
r_6	1
r_2	1
r_4	0
r_5	0
r_7	0

Table 18.
$C_1 \bowtie C_3$.

R	$C_2 \bowtie C_4$
r_1	0
r_3	0
r_6	0
r_2	1
r_4	1
r_5	1
r_7	1

Table 19.
$C_2 \bowtie C_4$.

R	C_2 U C_3
r_1	1
r_3	1
r_6	1
r_2	1
r_4	1
r_5	1
r_7	1

Table 20.
$C_2 \bowtie C_3$.

theoretical approach and the parallel cluster design through response vector approach.

The C-R cluster property between cluster set C and dataset R can be stated in terms of the properties of vectors. The data cluster incidences of cluster set C with C-R cluster property may be represented as response vector set V. For instance the cluster set $\{C_1, C_2, C_3, C_4\}$ has response vector set $\{V_1$=(1,1,1,0,0,0,0), V_2=(0,0,0,1,1,1,1), V_3=(1,1,1,0,0,0), and V_4=(0,0,0,0,1,1,1)$\}$ (**Tables 21–23**).

For instance, the response vector of the cluster $C1$ is given by column vector (1,1,1,0,0,0,0).

Suppose C_i and C_j are two clusters. If the two vectors V_i and V_j of C_i and C_j and the intersection $V_i \cap V_j = \Phi$, then the cluster set $\{C_i, C_j\}$ has a parallel cluster

R	C_1	C_2
r_1	1	0
r_3	1	0
r_6	1	0
r_2	0	1
r_4	0	1
r_5	0	1
r_7	0	1

Table 21.
$\{C_1, C_2\}$.

R	C3	C_4
r_1	1	0
r_3	1	0
r_6	1	0
r_2	1	0
r_4	0	1
r_5	0	1
r_7	0	1

Table 22.
$\{C_3, C_4\}$.

R	C_2	C_3
r_1	0	1
r_3	0	1
r_6	0	1
r_2	1	1
r_4	1	0
r_5	1	0
r_7	1	0

Table 23.
$\{C_2, C_3\}$.

property. Consider the vectors V_1 and V_2 of C_1 and C_2. The intersection of $V_1 \cap V_2 = \Phi$, so that the cluster set $\{C_1, C_2\}$ has parallel cluster property. Similarly the cluster set $\{C_3, C_4\}$ has parallel cluster property. The cluster set $\{C_2, C_3\}$ does not have parallel cluster property because $V_1 \cap V_2 \# \Phi$ and r_2 depending on C_1 and C_2.

3.2 Visual design for parallel cluster

The C-R cluster property is studied with graphical approach. This graphical approach can be studied for designing parallel cluster processing (PCP).

Suppose V_i is the vertex of RICM of C. The G(C) is defined by vertices V_i, i=1,2, ... , and n, and two vertices have an edge E_{ij} associated with interval $I_i=\{V_i, V_{i+1}\}$ i=1, ... ,n-1.

If G(C) has C-R cluster property, the vertices of G(C) have consecutive 1's or 0's.

Consider the cluster set $\{C_1, C_2\}$. The G(C1) has the vertices (1,1,1,0,0,0,0), and the G(C_2) has the vertices (0,0,0,1,1,1,1), G(C_3) has the vertices (1,1,1,1, 0,0,0), and G(C_4) has vertices (0,0,0,0,1,1,1).

The parallel cluster property exists if $G(C_i) \cap G(C_j)=\Phi$.

For instance, consider the G(C_1) and G(C_2). $G(C_1) \cap G(C_2)=\Phi$, so that the cluster set $\{C_1, C_2\}$ has parallel cluster property. The graphical representation is shown in **Figure 1**.

Similarly the cluster set $\{C_3, C_4\}$ has the parallel cluster property (PCP). The cluster set $\{C_3, C_4\}$ has no PCP because it is $G(C_2) \cap G(C_3)$ # Φ

The graph $G(C_1) \cap G(C_2) = \Phi$ have consecutive cluster property.

The graph $G(C_3) \cap G(C_4) = \Phi$ have consecutive cluster property. The graphical representation is shown in **Figure 2**.

The graph $G(C_2) \cap G(C_3)$ # Φ does not have consecutive cluster property. The graphical representation is shown in **Figure 3**.

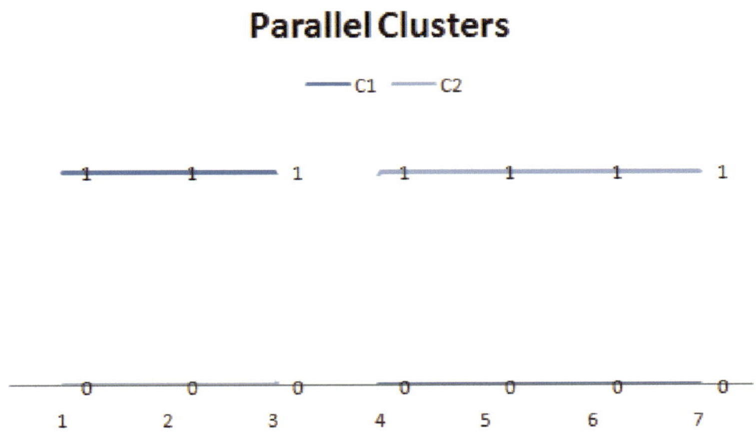

Figure 1.
{C₁, C₂}.

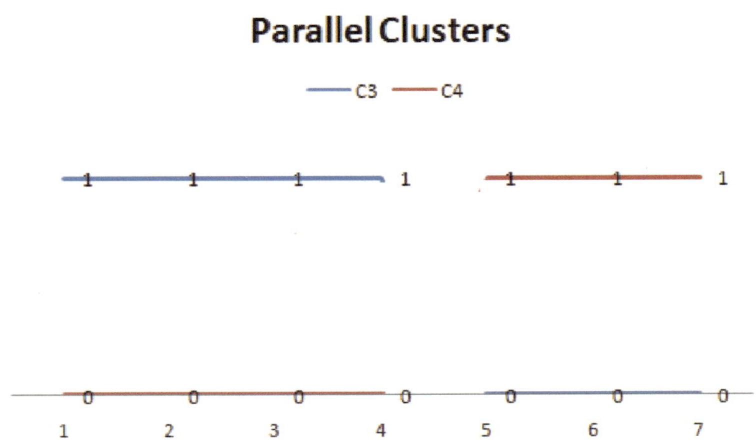

Figure 2.
$\{C_3, C_4\}$.

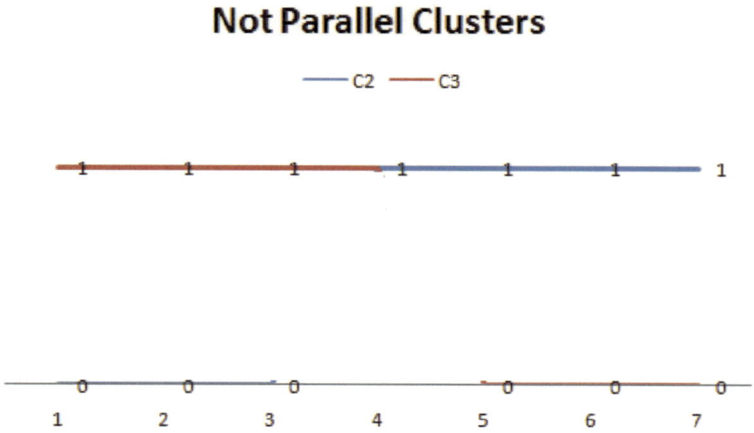

Figure 3.
$\{C_2, C_3\}$.

3.3 Parallel cluster design through genetic approach

Genetic algorithms (GAs) were introduced by Darwin [6]. GAs are used to learn and optimize the problem [7]. There are four evaluation processes:

- Selection
- Reproduction
- Mutation
- Competition

Consider the following crossover with two cuts:
Parent #1 00001111
Parent #2 11110000
The parent #1 and #2 match with crossover.
The C-R cluster property is studied through genetical study. This study will help for designing parallel cluster processing (PCP).
Definition: The gene G of cluster G(C) is defined as incidence sequence.
Suppose G(C_1) is parent and G(C_2) child genome of cluster incidence for C_1 and C_2.
Suppose the G(C_1) has (1,1,1,0,0,0,0) and the G(C_2) has the v(0,0,0,1,1,1,1).
The parallel cluster property may be designed using genetic approach with the C-R cluster property.
Suppose C is cluster set, R is dataset and G(C) is genetic set.
The parallel cluster property exists if G(C_i) and G(C_j) matches with crossover.
For instance,
G(C_1) = 11110000
G(C_2) = 00001111
G(C_1) and G(C_2)matches with the crossover.
The cluster set $\{C_1, C_2\}$ has parallel cluster property.
Similarly the cluster set $\{C_3, C_4\}$ has the parallel cluster property. The cluster set $\{C_3, C_4\}$ has no PCP because G(C_2) and G(C_3) are not matched with crossover.

3.4 Parallel cluster design cluster analysis

Clustering is grouping the particular data according to their properties, and sample clusters C_1 and C_2 are given in **Tables 24** and **25**, respectively.

R	C_1
r_1	1
r_3	1
r_6	1

Table 24.
Cluster C_1.

R	C_2
r_2	1
r_4	1
r_5	1
r_7	1

Table 25.
Cluster C_2.

Thus, the C_1 and C_2 have consecutive parallel cluster property (**Tables 26** and **27**).

R	C_3
r_1	1
r_3	1
r_5	1
r_6	1

Table 26.
Cluster C_3.

R	C_4
r_2	1
r_4	1
r_7	1

Table 27.
Cluster C_4.

Thus, the C_3 and C_4 have consecutive parallel properly. C_2 and C_3 do not have consecutive parallel cluster property because r_2 is common.

4. Design of retrieval of cluster using blackboard system

Retrieval of clusters from blackboard system [8] is the direct retrieval of data sources. When the query is being processed, the entire database has to bring to main

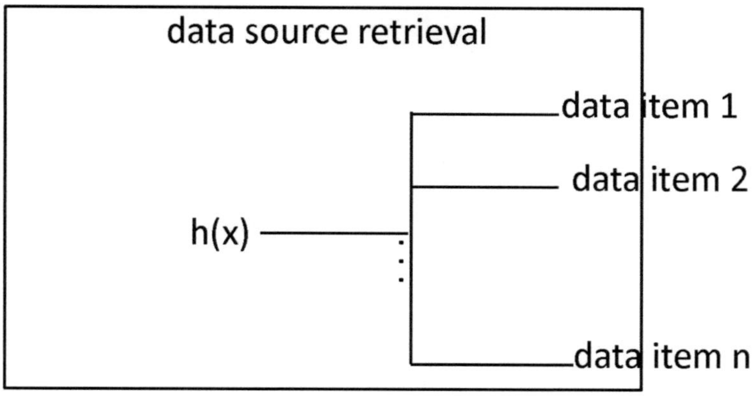

Figure 4.
Blackboard system.

memory but in blackboard architecture, the data item source is direct from the blackboard structure. For the retrieval of information for a query, data item is directly retrieved from the blackboard which contains data item sources. Hash function may be used to store the data item set in the blackboard.

The blackboard systems may be constructed with data structure for data item sources.

Consider the account (AC-No, AC-Name, AC-Balance)

Here AC-No is key of datasets.

Each data item is data sourced which is mapped by $h(x)$.

These data items are stored in blackboard structure.

When the transaction is being processed, there is no need to take the entire database into the main memory. It is sufficient to retrieval of particular data item of particular transaction from the blackboard system (**Figure 4**).

The advantage of blackboard architecture is highly secured for blockchain transaction. The blockchain technology has no third-party interference.

5. Fuzzy data mining

Sometimes, data mining is unable to deal with incomplete database and unable to combine the data and reasoning. Fuzzy data mining [6, 7, 9–18] will combine the data and reasoning by defining with fuzziness. The fuzzy MapReducing algorithms have two functions: *mapping* reads fuzzy datasets and *reducing* writes the after operations.

Definition: Given some universe of discourse X, a fuzzy set is defined as a pair $\{t, \mu_d(t)\}$, where t is tuples and d is domains and membership function $\mu_d(x)$ is taking values on the unit interval [0,1], i.e., $\mu_d(t) \rightarrow [0,1]$, where $t_i \in X$ is tuples (**Table 28**).

R1	d_1	d_2	.	d_m	μ
t_1	a_{11}	a_{12}	.	a_{1m}	$\mu_d(t_1)$
t_2	a_{21}	a_{22}		A_{2m}	$\mu_d(t_2)$
.	.	.	.	.	.
t_n	a_{1n}	a_{1n}	.	A_{nm}	$\mu_d(t_n)$

Table 28.
Fuzzy dataset.

The sale is defined intermittently with fuzziness (**Tables 29–32**).

CNo	INo	IName	Demand
C001	I005	shirt	0.9
C001	I007	Dress	0.65
C003	I004	pants	0.85
C002	I007	dress	0.6
C001	I008	Jacket	0.65
C002	I005	shirt	0.9

Table 29.
Fuzzy demand.

$\mu_{\text{Demand}}(x)$=0.9/90+0.85/80+0.8/75+0.65/70
or
Fuzziness may be defined with function
$\mu_{\text{Demand}}(x)$= $(1+(\text{Demand}-100)/100)^{-1}$ Demand <=100
=1 Demand>100

A. *Negation*

CNo	INo	IName	Negation of price
C001	I005	shirt	0.3
C001	I007	Dress	0.5
C003	I004	pants	0.4
C002	I007	dress	0.5
C001	I008	Jacket	0.4
C002	I005	shirt	0.3

Table 30.
Negation of price.

A. *Union*

CNo	INo	IName	Sales U price
C001	I005	Shirt	0.8
C001	I007	Dress	0.5
C003	I004	Pants	0.6
C002	I007	Dress	0.5
C001	I008	Jacket	0.6
C002	I005	Shirt	0.7

Table 31.
Sales U price.

Union of 1105 = max{0.8,0.7}=0,8
Fuzzy semijoin is given by sales ⋈ items-sale as shown in **Table 33**.

INo	IName	Sales
I005	Shirt	0.8
I007	Dress	0.5
I004	Pants	0.6
I007	Dress	0.5
I008	Jacket	0.6

Table 32.
Items-sales.

CNo	INo	IName	Sales
C001	I005	shirt	0.8
C001	I007	Dress	0.5
C003	I004	pants	0.6
C002	I007	dress	0.5
C001	I008	Jacket	0.7
C002	I005	shirt	0.7

Table 33.
Fuzzy semijoin.

The fuzzy k-means clustering algorithm (FKCA) is optimization algorithm for fuzzy datasets (**Table 34**).

CNo	INo	IName	Sales
C001	I005⇔I007	Shirt⇔Dress	0.4
C003	I004	pants	0.6
C002	I007⇔I005	Dress⇔shirt	0.5

Table 34.
Association.

Fuzzy k-means cluster algorithm (FKAC) is given by, using FAD
best=R
K=means=best
for *i* range(1,*n*)
  for *j* range(1,*n*)
    t_i=fuzzy union(r_i.RU r_i.R_j), if r_i.R=r_j.R
C reduce best
k-means < best
return

The fuzzy multivalued association property of data mining may be defined with multivalued fuzzy functional dependency.

The fuzzy multivalued association (FMVD) is the multivalve dependency (MVD). The association multivalve dependency (FAMVD) may be defined by using Mamdani fuzzy conditional inference [3].

If $EQ(t_1(X),t_2(X),t_3(X))$ then $EQ(t_1(Y),t_2(Y))$ or $EQ(t_2(Y),t_3(Y))$ or $EQ(t_1(Y), t_3(Y))$

$= \min\{EQ(t_1(Y),t_2(Y))\ EQ(t_2(Y),t_3(Y))\ EQ(t_1(Y),t_3(Y))\}$

$= \min\{\min(t_1(Y),t_2(Y)),\ \min(t_2(Y),t_3(Y)),\ \min(t_1(Y),t_3(Y))\}$

$= \min(t_1(Y), t_2(Y). t_3(Y))$

The fuzzy k-means clustering algorithm (FKCA) is the optimization algorithm for fuzzy datasets (**Table 35**).

CNo	INo	IName	Sales
C001	I005⇔I007 ⇔I008	Shirt⇔Dress ⇔Jacket	0.8 0.4 0.5
C003	I004	Pants	0.6
C002	I007⇔I005	Dress⇔shirt	0.5 0.7

Table 35.
Association using AFMVD.

Fuzzy k-means cluster algorithm (FKAC) is given by, using FAMVD
best=R
K=means=best
for i range(1,n)
  for j range(1,n)
    for k range(1,n)
      t_i=fuzzy union($r_i.R$ U $r_j.R$ U $r_k.R$), if $r_i.R=r_j.R=r_k.R$
C reduce best
k-means<best
return

The fuzzy k-means clustering algorithm (FKCA) is the optimization algorithm for fuzzy datasets.
K=means=n
for i range(1,n)
  for j range(1,n)
    t_i=fuzzy union(r_i.R U $s_i.S_j$), if $r_i.R=s_j.S$
C =best
k-means < best
return

For example, consider the sorted fuzzy sets of **Table 5** is given in **Table 36**.

CNo	INo	IName	Sales ⋈ Price⋈ Demand
C001	I005	Shirt	0.8
C001	I007	Dress	0.5
C003	I004	Pants	0.6
C002	I007	Dress	0.5
C001	I008	Jacket	0.6
C002	I005	Shirt	0.7

Table 36.
Fuzzy join.

6. Fuzzy security for data mining

Security methods like encryption and decryption are used cryptographically. These security methods are not secured. Fuzzy security method is based on the mind and others do not descript. Zadeh [16] discussed about web intelligence, world knowledge, and fuzzy logic. The current programming is unable to deal question answering containing approximate information. For instance "which is the best car?" The fuzzy data mining with security is knowledge discovery process with data associated.

The fuzzy relational databases may be with fuzzy set theory. Fuzzy set theory is another approach to approximate information. The security may be provided by approximate information.

Definition: Given some universe of discourse X, a relational database $R1$ is defined as pair $\{t, d\}$, where t is tuple and d is domain (**Table 37**).

R1	d_1	d_2	.	d_m
t_1	a_{11}	a_{12}	.	a_{1m}
t_2	a_{21}	a_{22}		A_{2m}
.	.	.	.	.
t_n	a_{1n}	a_{1n}	.	A_{nm}

Table 37.
Relational database.

Price = 0.4/50+0.5/60+07/80+0.8/100

The fuzzy security database of price is given in **Table 38**.

INo	IName	Price
I005	Benz	0.8
I007	Suzuki	0.4
I004	Toyota	0.7
I008	Skoda	0.5
I009	Benz	0.8

Table 38.
Price fuzzy set.

Demand = 0.4/50+0.5/60+0.7/80+0.8/100

The fuzzy security database of demand is given in **Table 39**.

INo	IName	Demand	μ
I005	Benz	80	0.7
I007	Suzuki	60	0.5
I004	Toyota	100	0.8
I008	Skoda	50	0.4
I009	Benz	80	0.7

Table 39.
Demand fuzzy set.

The lossless natural join of demand and price is union and is given in **Table 40**.

ino	Iname	Demand	price	μ
I005	Benz	80	100	0.8
I007	Suzuki	60	50	0.5
I004	Toyota	100	80	0.8
I008	Skoda	50	60	0.5
I009	Benz	80	100	0.8

Table 40.
Lossless join.

The actual data has to be disclosed for analysis on the web. There is no need to disclose the data if the data is inherently define with fuzziness.

"car with fuzziness >07" may defined as follows:

For instance,
XML data may be defined as
<CAR>
<COMPANY>
<NAME> Benz <NAME>
<FUZZ> 0.8 <FUZZ>
</COMPANY>
<COMPANY>
<NAME> Suzuki <NAME>
<FUZZ> 0.9<FUZZ>
</COMPANY>
<COMPANY>
<NAME> Toyoto<NAME>
<FUZZ> 0.6<FUZZ>
</COMPANY>
<COMPANY>
I<NAME> Skoda<NAME>
<FUZZ> 0.7<FUZZ>
</COMPANY>

Xquery may define using projection operator for demand car is given as
Name space default = http:\www.automoble.com/company
Validate <CAR> {
For $name in COMPANY/CAR
where $company/ Max($demand>0.7)}
return <COMPANY> {$company/name, $company/fuzzy}</COMPANY>
</CAR>

The fuzzy reasoning may be applied for fuzzy data mining.

Consider the more demand fuzzy database by decomposition (**Tables 41** and **42**).

The fuzzy reasoning [14] may be performed using Zadeh fuzzy conditional inference

The Zadeh [14] fuzzy conditional inference is given by
if x is P_1 and x is P_2 .... x is P_n then x is Q =
$\min 1, \{1-\min(\mu_{P1}(x), \mu_{P2}(x), \ldots, \mu_{Pn}(x)) + \mu_Q(x)\}$

INo	IName	Demand
I005	Benz	0.8
I007	Suzuki	0.9
I004	Toyota	0.6
I008	Skoda	0.7
I009	Benz	0.9

Table 41.
Demand.

INo	IName	Price
I005	Benz	0.7
I007	Suzuki	0.4
I004	Toyota	0.6
I008	Skoda	0.5
I009	Benz	0.7

Table 42.
Price.

The Mamdani [7] fuzzy conditional inference s given by
if x is P_1 and x is P_2 …. x is P_n then x is Q =
min $\{\mu_{P1}(x), \mu_{P2}(x), \ldots, \mu_{Pn}(x), \mu_Q(x)\}$
The Reddy [12] fuzzy conditional inference s given by
$= \min(\mu_{P1}(x), \mu_{P2}(x), \ldots, \mu_{Pn}(x))$
If x is Demand then x is price
x is more demand

---

x is more Demand o (Demand → Price)
x is more Demand o min{1, 1-Demand+Price}Zadeh
x is more Demand o min{Demand, Price} Mamdani
x is more Demand o {Demand} Reddy
"If x is more demand, then x is more prices" is given in **Tables 43** and **44**.
The inference for price is given in **Table 45**.
So the business administrator (DA) can take decision to increase the price or not.

INo	IName	More demand
I005	Benz	0.89
I007	Suzuki	0.95
I004	Toyota	0.77
I008	Skoda	0.84
I009	Benz	0.95

Table 43.
More demand.

INo	IName	Zadeh	Mamdani	Reddy
I005	Benz	0.9	0.7	0.7
I007	Suzuki	0.5	0.4	0.4
I004	Toyota	1,0	0.6	0.6
I008	Skoda	0.8	0.5	0.5
I009	Benz	0.8	0.7	0.7

Table 44.
Demand → Price.

INo	IName	Zadeh	Mamdani	Reddy
I005	Benz	0.89	0.7	0.7
I007	Suzuki	0.5	0.4	0.4
I004	Toyota	0.77	0.6	0.6
I008	Skoda	0.8	0.5	0.5
I009	Benz	0.8	0.7	0.7

Table 45.
Inference price.

7. Web intelligence and fuzzy data mining

Let C and D be the fuzzy rough sets (**Tables 46–51**).

	d_1	2_2	.	d_m	μ
t_1	a_{11}	a_{12}	.	a_{1m}	$\mu_d(t_1)$
t_2	a_{21}	a_{22}		A_{2m}	$\mu_d(t_2)$
.	.	.	.	.	.
t_n	a_{1n}	a_{1n}	.	A_{nm}	$\mu_d(t_n)$

Table 46.
Fuzzy database.

INo	IName	Price	μ
I005	Shirt	100	0.8
I007	Dress	50	0.4
I004	Pants	80	0.7
I008	Jacket	60	0.5
I009	Skirt	100	0.8

Table 47.
Price database.

The operations on fuzzy rough set type 2 are given as
$1\text{-}C= 1\text{-}\ \mu_C(x)$ Negation
$C\vee D=\max\{\mu_C(x), \mu_D(x)\}$ Union
$C\wedge D=\min\{\mu_C(x) , \mu_D(x)\}$ Intersection

INo	IName	Demand	Price	μ
I005	Shirt	80	100	0.7
I007	Dress	60	50	0.4
I004	Pants	100	80	0.7
I008	Jacket	50	60	0.4
I009	Skirt	80	100	0.7

Table 48.
Intersect of demand and price.

INo	IName	Demand	μ
I005	Shirt	80	0.8
I007	Dress	60	0.5
I004	Pants	100	0.8
I008	Jacket	50	0.5
I009	Skirt	80	0.8

Table 49.
Lossless decomposition of demand.

INo	IName	Price	μ
I005	Shirt	100	0.8
I007	Dress	50	0.5
I004	Pants	80	0.8
I108	Jacket	60	0.5
I009	Skirt	100	0.8

Table 50.
Lossless decomposition of price.

Company	μ
IBM	0.8
Microsoft	0.9
Google	0.75

Table 51.
Best software company.

XML data may be defined as
<SOFTWARE>
<COMPANY>
<NAME> IBM <NAME>
<FUZZ> 0.8 <FUZZ>
</COMPANY>

<COMPANY>
<NAME> Microsoft <NAME>
<FUZZ> 0.9<FUZZ>
</COMPANY>
<COMPANY>
<NAME> Google<NAME>
<FUZZ> 0.75<FUZZ>
</COMPANY>

Xquery may define using projection operator for best software company is given as

Name space default = http:\www.software.cm/company

Validate <SOFTWARE> {For $name in COMPANY/SOFTWARE where $company/ Max($fuzz)}

return <COMPANY> {$company/name, $company/fuzzy} </COMPANY>

</SOFTWARE>

Similarly, the following problem may be considered for web programming.

Let P is the fuzzy proposition in question-answering system.

P=Which is tallest buildings City?

The answer is "x is the tallest buildings city."

For instance, the fuzzy set "most tallest buildings city" may defined as

most tallest buildings city = 0.6/Hoang-Kang + 0.6/Dubai + 0.7/New York +0.8/ Taipei+ 0.5/Tokyo

For the above question, output is "tallest buildings city"= 0.8/Taipei by using projection.

The fuzzy algorithm using FUZZYALGOL is given as follows:

BEGIN

Variable most tallest buildings City = 0.6 / Hoang-Kang + 0.6 / Dubai + 0.7 / New York + 0.8 / Taipei + 0.5 / Tokyo

most tallest buildings City =0.8 / Taipei

Return URL, fuzziness=Taipei, 0.8

END

The problem is to find "most pdf of type-2 in fuzzy sets"

The Fuzzy algorithm is

Go to most visited fuzzy set cites

Go to most visited fuzzy sets type-2

Go to most visited fuzzy sets type -2 pdf

The web programming gets "the most visited fuzzy sets" and put in order

The web programming than gets "the most visited type-2 in fuzzy sets"

The web programming gets "the most visited pdf in type-2"

8. Conclusion

Data mining may deal with incomplete information. Bayesian theory needs exponential complexity to combine data. Defining datasets with fuzziness inherently reduce complexity. In this chapter, fuzzy MapReduce algorithms are studied based on functional dependencies. The fuzzy k-means MapReduce algorithm is studied using fuzzy functional dependencies. Data mining and fuzzy data mining are discussed. A brief overview on the work on business intelligence is given as an example.

Most of the current web programming studies are unable to deal with incomplete information. In this chapter, the web intelligence system is discussed for fuzzy data mining. In addition, the fuzzy algorithmic language is discussed for design fuzzy algorithms for data mining. Web intelligence system for data mining is discussed. Some examples are given for web intelligence and fuzzy data mining.

Acknowledgements

The author thanks the reviewer and editor for revision and review suggestions made in this work.

Author details

Poli Venkata Subba Reddy
Department of Computer Science and Engineering, Sri Venkateswara University, Tirupati, India

*Address all correspondence to: pvsreddy@hotmail.co.in

References

[1] Ghosh SP. File organization: The consecutive retrieval property. Communications of the ACM. 1972; **15**(9):802-808

[2] Chin FY. Effective Inference Control for Range SUM Queries, Theoretical Computer Science, 32,77-86. North-Holland; 1974

[3] Kamber M, Pei J. Data Mining: Concepts and Techniques. New Delhi: Morgan Kaufmann; 2006

[4] Ramakrishnan R, Gehrike J. Data Sets Management Systems. New Delhi: McGraw-Hill; 2003

[5] Tan PN, Steinbach V, Kumar V. Introduction to Data Mining. New Delhi: Addison-Wesle; 2006

[6] Zadeh LA. Fuzzy logic. In: IEEE Computer. 1988. pp. 83-92

[7] Tanaka K, Mizumoto M. Fuzzy programs and their executions. In: Zadeh LA, King-Sun FU, Tanaka K, Shimura M, editors. Fuzzy Sets and Their Applications to Cognitive and Decision Processes. New York: Academic Press; 1975. pp. 47-76

[8] Englemore R, Morgan T. Blackboard Systems. New Delhi: Addison-Wesley; 1988

[9] Poli VSR. On existence of C-R property. Proceedings of the Mathematical Society. 1989;**5**:167-171

[10] Venkta Subba Reddy P. Fuzzy MapReduce Data Mining Algorithms, 2018 International Conference on Fuzzy Theory and Its Applications (iFUZZY2018), November 14-17; 2108

[11] Reddy PVS, Babu MS. Some methods of reasoning for conditional propositions. Fuzzy Sets and Systems. 1992;**52**(3):229-250

[12] Venkata Subba Reddy P. Fuzzy data mining and web intelligence. In: International Conference on Fuzzy Theory and Its Applications (iFUZZY); 2015. pp. 74-79

[13] Reddy PVS. Fuzzy logic based on belief and disbelief membership functions. Fuzzy Information and Engineering. 2017;**9**(9):405-422

[14] Zadeh LA. A note on web intelligence, world knowledge and fuzzy logic. Data and Knowledge Engineering. 2004;**50**:91-304

[15] Zadeh LA. A note on web intelligence, world knowledge and fuzzy logic. Data and Knowledge Engineering. 2004;**50**:291-304

[16] Zadeh LA. Calculus of fuzzy restrictions. In: Zadeh LA, King-Sun FU, Tanaka K, Shimura M, editors. Fuzzy Sets and Their Applications to Cognitive and Decision Processes. New York: Academic Press; 1975. pp. 1-40

[17] Zadeh LA. Fuzzy algorithms. Information and Control. 1968;**12**: 94-104

[18] Zadeh LA. Precipitated Natural Language (PNL). AI Magazine. 2004; **25**(3):74-91